Control Loop Foundation
Batch and Continuous Processes

Robert,
Thanks for attending this process
introduction class.

Terry Blevins

Control Loop Foundation

Batch and Continuous Processes

Terrence Blevins
Mark Nixon

Copyright © 2011   International Society of Automation
All rights reserved.   67 Alexander Drive
                          P.O. Box 12277
                          Research Triangle Park, NC 27709

Printed in the United States of America.
10 9 8 7 6 5 4 3

ISBN 978-1-936007-54-7

Library of Congress Cataloging-in-Publication Data is in process.

Dedication

This book is dedicated to Karen Blevins and Nancy Nixon, who have provided encouragement and support throughout our careers.

Acknowledgments

The authors wish to express their appreciation to Grant Wilson for supporting work on this book and to Duncan Schleiss, John Berra, Bud Keyes, Ron Eddie, Tom Aneweer, Steve Boyce, Darrin Kuchle, John Caldwell, Dennis Stevenson, Willy Wojsznis, Mike Sheldon, David Holmes, Bob Lenich, Tim Prickette, Dave Deitz, Craig Sydney, Jim Siemers, Mark Dimmitt, Jim Hoffmaster and Gil Pareja from Emerson Process Management for their inspiration and support of process control initiatives. In our work, we have benefited from communications with Karl Åström from Lund University, Tom Edgar from the University of Texas, and Dale Seborg from the University of California, Santa Barbara on both basic and advanced control topics. The authors wish to thank Jim Cahill and Deborah Franke for their guidance in the design of the web site for this book and Brenda Forsythe and Jim Sipowicz for the creative book cover design. We want to thank Karen Blevins, Mark Sowell, Professor Tom Edgar, Greg McMillan, and Scott Bogue for their review inputs, and to thank Susan Colwell, Manager, Publications Development, ISA, for her support in the publication of this book.

Over the years, we have benefited from working with many others in the design and implementation of control systems. It was an honor to work with Greg Graziadio, Puffer Sweiven Inc.; his influence can be seen in some of the application examples included in this book. Our early work on control projects and the development of control techniques for ammonia and pulp and paper processes was encouraged by Bruce Duncan, Sheldon Lloyd, Chuck Schuder, Charlie Brez, Bob Otto, John Hedrick, Tom Bell, Sid Smith, Rick Genter, Ken Langley, and Dick Seemann of Emerson Process Management. We also gratefully acknowledge the support of the many customers we have worked with on control projects. In particular, the following individuals supported the pursuit of new control applications and control technologies: Harry Pinder, Chris Liakos, Roger Smith, Bobby Deaton, Howard Lane, Mike Donohoe, George Wallace, and Roger Nesbit, Georgia Pacific; Eric Striter, GE; Paul Friedman, Allied Chemical; David Taylor, Dow; Howard Bickley, Union Camp; James Beall and John Traylor, Texas Eastman; Robert Chamberlain and Bob Michelson, MacMillan Bloedel Limited; George Fink, US Borax; Mark Sowell and Greg McMillan, Solutia; Romeo Ancheta, Husky Energy; Derrick Vanderkraats, Canfor; Bruce Johnson and Efren Hernandez, Lubrizol; and Bruce Eldridge, Frank Seibert, and Robert Montgomery, University of Texas, Austin, Pickle Research Center.

It has been gratifying to work with Terry Chmelyk, Saul Mtakula, and the rest of Norpac's control group in field testing new control technologies. The lime kiln example in this book reflects some of their work in this area. We greatly appreciate the support of Dave Wall, Don Umbach, John Peterson, and the rest of the Norpac management team and their passion for process control. We also wish to thank Mike Begin and the folks at Spartan Controls; Dan Moody and Harley Jeffery, ControlSouthern; and Randy Angelle, John H. Carter Co. for their continued support throughout our careers.

About the Authors

Terrence "Terry" Blevins has been actively involved in the application and design of process control systems throughout his career. For more than 15 years, he worked as a systems engineer and group manager in the design and startup of advanced control solutions for the pulp and paper industry. Terry was instrumental in the establishment of Emerson Process Management's Advanced Control Program. From 1998–2005, Terry was the team lead for the development of DeltaV advanced control products. He is the Fieldbus Foundation™ team lead for the development and maintenance of the Function Block Specification and editor of the SIS Architecture and Model Specifications. In this capacity, Terry is involved in the movement of Fieldbus Foundation function block work into international standards. Terry is the U.S. expert to the IEC SC65E WG7 function block committee that is responsible for the IEC 61804 function block standards. He is a voting member and chairman of ISA104-EDDL (Electronic Device Description Language) committee and is the technical advisor to the United States Technical Advisory Group (USTAG) for the IEC65E subcommittee. He is also a member of the USNC TAG (IEC/SC65 and IEC/TC65). Terry authored "An Overview of the ISA/IEC Fieldbus," Section 11, Standards Overview, Fifth Edition of the *Process/Industrial Instruments and Controls Handbook* and coauthored four sections in the Fourth Edition of the *Instrumentation Engineer's Handbook, Process Control and Optimization*. He coauthored the ISA bestselling book *Advanced Control Unleashed*. He has 36 patents and has written over 65 papers on process control system design and applications. Terry received a Bachelor of Science in Electrical Engineering from the University of Louisville in 1971 and a Master of Science in Electrical Engineering from Purdue University in 1973. In 2004, he was inducted into *Control Magazine*'s Process Automation Hall of Fame. Presently, Terry is a principal technologist in the future architecture team of DeltaV Product Engineering at Emerson Process Management.

Phone: (512) 418-4628 E-mail: terry.blevins@emerson.com

Mark Nixon has been involved in the design and development of control systems throughout his career. Mark started his career as a systems engineer working on projects in oil & gas, refining, chemicals, and pulp & paper. He moved from Canada to Austin, TX in 1998 where he has held a variety of positions in both research and development. From 1995–2005, Mark was lead architect for DeltaV. In 2006, he joined the wireless team, taking a very active role in the development of the

WirelessHART specifications and the development of the IEC 62591 standardization. Mark's current research includes control using WirelessHART devices, data analytics for batch process, use of wireless technology the process industry, mobile users, operator interfaces, and advanced graphics. He is currently active in the Center for Operator Performance (http://www.operatorperformance.org), WirelessHART, ISA-88 standard, Foundation Fieldbus standards (http://www.fieldbus.org/), and ISA-101 standard. He has written numerous papers and currently holds more than 45 patents. He coauthored *WirelessHART: Real-Time Mesh Network for Industrial Automation* and has made contributions to the *Industrial Instruments and Controls Handbook* and *Essentials of Modern Measurements and Final Elements in the Process Industry*. Mark received his Bachelor of Science in Electrical Engineering from the University of Waterloo in 1982.

Phone: (512) 418-7445 E-mail: mark.nixon@emerson.com

Contents

Foreword

Today technical staff at a modern process plant often face a bewildering array of process equipment, sensors, control valves, computer hardware and software packages, yet with all these complexities, it is important to run the plant so that it achieves profitability, satisfies environmental regulations, minimizes energy consumption, and avoids hazardous situations. The effective operation of computer-based control systems is critical to realize the above objectives; thus, it is important to have well-trained engineers and technicians to support such systems. *Control Loop Foundation – Batch and Continuous Processes* has been authored by Terry Blevins and Mark Nixon of Emerson Process Management to provide staff in operating plants with introductory training that does not require a high level of mathematics and simulation knowledge. I read this book with great interest because I have known both of the authors since the mid-1990s, and my students and I have interacted with them in several process control research projects. In addition, I have co-authored several editions of the leading process control textbook, *Process Dynamics and Control*, Dale E. Seborg, Thomas F. Edgar, and Duncan A. Mellichamp during that same period of time. It is remarkable that *Control Loop Foundation – Batch and Continuous Processes* so ably complements the textbook used in many university chemical engineering departments. Terry Blevins and Mark Nixon's book is written at such a level that plant operations and technical personnel can easily relate to the content because of its practical orientation. The book will also help plant technical staff to integrate the many disparate pieces of information they know about process control loops so that they can have a unified view of instrumentation, field communications, control strategies, process dynamics, loop tuning and performance, complex control systems, and practical control applications. The style of the material's presentation is well-adapted to individual study, and the authors have also included workshops and application examples that readers with web access may interactively work with to develop an intuitive feel for the dynamic behavior of typical control loops. I congratulate the authors on this impressive achievement and believe it will have considerable impact in those operating plants where it is used.

Thomas F. Edgar
University of Texas - Austin

1

Introduction

This book originally started as a special class for new engineers within one of Emerson Process Management's engineering divisions, but has since grown in both scope and depth of material that is addressed. There are many aspects of process control systems, and the book is structured to allow engineers, managers, technicians, and others that are new to process control to get up to speed more quickly on process control and related areas. Experienced control engineers will benefit from the application examples and workshops on process control design and implementation of multi-loop control strategies.

The material is presented in a manner that is independent of the control system manufacturer.

The background material included in the first part of the book will be helpful to a new engineer who is just starting in this field and perhaps has never worked in a plant environment. Much of the material presented on the practical aspects of control system design and process applications is typically not included in process control taught at the university level.

Many of the topics that are addressed in the book are areas that the authors have learned through hands-on experience gained while working in the design and commissioning of process control systems. Also, we have benefited from the insight of many people working in process control. This is a good way to learn, but it's maybe not always the best or most efficient way to become proficient in these areas. Our goal, then, is to address concepts and terminology that an engineer needs when working in the process industry. We hope that this information will provide helpful insight as you look at a specific control system and how it's used, and how best to use this equipment while addressing the different application requirements. Whether you are working as a process control engineer in a manufacturing plant, working in a controls group in an engineering department, or working in an instrumentation department within a manufacturing plant, we hope the information provided here will help you in your work and will set a solid foundation that allows you to confidently address new control applications in the future.

A lot has changed in the process industry over the last 30 some-odd years since we first started work as process control engineers. When we first started working with process control systems, as an engineer you often had an opportunity to see the whole project—working with the plant, developing the controls strategy, documentation, and user interface, and then commissioning the system. In some cases, this involved modifying the control system software and hardware required to support the control strategy. This opportunity existed because at that time, the control groups were often small and were focused on supporting plant operations. In some cases, you'd be given a plane ticket, travel to the plant site, discuss the application's requirements, come back to the office to develop the process control strategy and user interface, and then return to the plant to commission the control system. The ultimate measure of success was whether the plant was happy with the system's performance and with the benefits realized as a result of improved plant efficiency or throughput. As an engineer, you also had the personal satisfaction of providing the operators with better tools to manage and improve plant performance.

Today, many control system design organizations are much bigger; thus, it is possible for a person working on some aspects of a control system not to be involved in operator training or the commissioning of the control system. This is unfortunate since in that case you don't have an opportunity to see the full picture of how things fit together or to get direct and immediate feedback on how well a new or updated control system performed and whether it was necessary to make changes in the field to meet plant operators' or management's operational requirements. So, the application examples in each chapter are designed to help you gain an appreciation for all aspects of the design, implementation, and commissioning of a control system.

The book's background material is organized so that new concepts build on material presented in previous sections. As you read the book, we suggest you cover the material in the order it is presented. If you have worked with control systems in the past, then the first part of the book may cover material you already know. Even so, it would be a good idea to at least review those chapters. Many of the terms introduced in the first portion of the book are defined to establish a basis for understanding, which will be required to appreciate and apply the material on control system design and implementation presented in later sections of the book. Also, this background material is intended to promote an appreciation for the way in which many of the terms and concepts that form the foundation of industrial process control have evolved over time and through the efforts of many people involved in control system design and implementation. An understanding of these terms and why they have been traditionally described in a certain way can be helpful when considering the best way to meet the requirements of a new process. Also, with this understanding, it should become clear that the control system can and should be designed independent of the equipment manufacturer and the technology used.

We also take the approach that you may not have worked in a process plant environment and that you may be unfamiliar with the field devices that the control system interfaces with, how these devices work, and the limitations of these devices. Thus, the background material in the first part of the book covers concepts and terminology that you'll find helpful in working with these field devices. For example, some chapters are dedicated to an overview of wiring and the field devices that are typically used in industry. The field devices (measuring devices and actuators) selected to instrument a process can impact control system performance, including the accuracy and speed with which changes in the process can be sensed and corrected. In addition, the wiring of the control system varies depending on whether a field device uses analog or digital technology.

Throughout the book, we show examples of drawings and other documents that are typical of what is used in industry to support control system design and maintenance. If you are involved in control system design, then it will be necessary for you to create documents that outline the basic control requirements and detail the measurement, calculation, and control strategy that will be used to address these requirements. Within the process industry, the engineering firm that initially designs a plant or the engineering department of a company that commissions new plant construction establishes the standards for control system documentation. In nearly all cases, the basic types of documents used in the process industry are fairly well standardized, even though there may be some variation at the detail level. Process examples are included to illustrate how these documents are used during the design and maintenance of the control system. An understanding of control system documentation is helpful when you are talking to someone about process control since such discussions often involve pulling out and reviewing some of these documents. Thus, it is important to understand why these documents are created, what information each document should communicate, and how to understand the symbols and terminology used in these documents.

Having established a background on field devices and control system documentation, we then address techniques that may be used to describe the dynamic behavior of a process. In the chapter on process characterization, we introduce terminology to describe both the static and dynamic responses of a process. This material sets a foundation for the remaining chapters on control design. How process behavior impacts control system design and commissioning will become clear as we address control system design.

At various points in the chapters on process characterization and control system design, you will have an opportunity to apply these concepts by using example workshops. These workshops are provided on a website and thus may be accessed without special tools or special software. The only thing that is required is access to a high-speed Internet connection and the web browser installed on your personal computer. In the Appendix, we provide detailed directions on how to access the website that supports these workshops. A video that shows the workshop solution is contained on the website. In addition, after some chapters there are questions that you may want to review and answer to judge how well you understand and remember the material. These questions are structured to be fun as well as informative. As you consider the appropriate answer to a question, feel free to go back through the chapter if you are uncertain of the answer. The objective of these study questions is to reinforce the learning process.

In looking at control system design, it is important to keep in mind the plant's requirements and how the control system will be used in the plant. For example, what are the production and quality objectives that must be met? If you don't have a clear picture of the goals to be achieved, it will be difficult to choose the measurement and control techniques that should be employed in the control system design. As part of the material presented on control system design, we'll review the primary function performed by a control system, and then present each of the control techniques that are most often used in industry today.

In general, the world of process control is broken into batch and continuous control. In a batch environment, where control is more sequence- or logic-oriented, the manufacturing steps are described by a procedure called a "recipe." A batch process is characterized by a constantly changing environment in which the controls and measurements must function over a wide range of operating conditions. A continuous control environment differs in that the objective is to maintain the process at the set of operating conditions needed to sustain a target production rate and quality level. In this book, we focus on the continuous control techniques that are the foundation of both batch and continuous applications. An overview is provided on the terminology used to describe batch procedures; thus, for more information on the structure of a batch recipe, the user is referred to the relevant standards, which are unique to the batch industry.

After covering the techniques that are traditionally used in industry, we then provide a brief introduction to advanced control. The material on advanced control techniques is designed to show how these tools may be used to better address the control requirements of interactive processes or processes characterized by difficult dynamics. A good understanding of traditional continuous control is needed to be able to evaluate whether these advanced control techniques are necessary to meet your control requirements. If you find this material on advanced control of interest, then there are a number of other references that address advanced control techniques in more detail.

A dynamic process simulation is built into the example workshops mentioned above, in order to give you an opportunity to obtain realistic "hands-on" experience with the control techniques presented in this book. The authors have found that during the pre-commissioning checkout of a control system, it is often helpful to create a dynamic process simulation that interacts with the control system. A process simulation environment that works with the control system may also be used to create an operator training program. When the process simulation is tied into the control system, the control operation and process response will approach those encountered in actual plant operation. A chapter is dedicated to the techniques that may be used to create a process simulation using the same tools that are commonly available within most process control systems.

Throughout the book, as control techniques are introduced, how each technique may be applied to address various process control requirements is illustrated using simple process examples. The last chapter of the book contains additional, more complex examples that illustrate how these basic techniques may be combined to meet a variety of control requirements. If you have read the preceding chapters and worked through the associated workshops, then you will have the foundation required to understand these more advanced process examples. As you study these examples, it should become clear that these more complex control techniques are made up of various combinations of the basic control techniques that we covered in previous chapters. It is our hope that the understanding you achieve by reading the book and going through the workshop exercises will set a foundation that will serve you well in addressing other control applications you may encounter.

In developing the book, the authors were careful to show how process control ties into the operator interface. Operators must be able to monitor the process and respond to changes in plant conditions. A significant part of the operator interface is directly tied to the control strategy. Through the operator interface, operators start and stop batch procedures, change setpoints, change production rates, start up and shut down equipment, and perform a wide range of other actions. The alarm system is often carefully designed to focus operator attention on the highest priority items first. As part of the design process, displays are carefully organized to show information related to a specific piece of equipment, a particular process, or some sequence of operations. Although the overall design of operator displays is beyond the scope of this book, enough information is provided to give you a good understanding of the working relationship between the control strategy and the operator graphic displays.

The reader should feel free to contact the authors at their email addresses if they have questions about the book or about the use of the web-based workshops. All royalties from this book will be given directly to universities and education programs to promote and enhance the understanding of process control. A beneficiary of each year's royalties will be chosen by the authors.

2

Background and Historical Perspective

This chapter provides a brief history of process control systems and an overview of the components that make up process plants. A quick survey of plants in the process industry would show that there is wide variation in the way that plants are physically designed and organized. There may be differences in plant layout, construction, and process equipment depending upon where a plant is located and the products that are produced by the plant. In this chapter, we provide an overview of process plants and the terminology that is commonly used in discussing the process, the process equipment, and the control system. Examples are used to illustrate the different ways a plant may be structured and the organization of process equipment within the plant. Some insights are provided by the authors into the importance of the plant operator in the plant operation. The manner in which off-line measurement of quality parameter can be made by labs within the plant is addressed. Also, the role of industry standards and specialized subsystems play in the design of modern control systems is detailed in this chapter.

2.1 Plant Structure

The physical location of the plant often dictates the approach that must be taken in plant design and the materials of construction. Plants are designed to meet some financial objective and where possible the construction cost and the raw material and transportation costs are minimized by selection of the plant site. For example, in the northern parts of the North American continent an enclosure may be built around some or all of the process equipment. In the pulp and paper mills located in Canada, it is quite common to build an enclosure around the process so that even the larger pieces of equipment, such as a continuous digester that's 300 feet tall, are not visible from outside the plant. The reason for this type of construction is that in these locations the outside air temperature may drop to about -30°F (-34.4°C) in the wintertime. Without the protection provided by the walls and insulation, the process equipment would be exposed to extremely low temperatures and this would adversely impact the efficiency and operation of the process.

In other industry segments, such as pharmaceutical manufacturing, where high value or highly regulated products may be produced, it is the products that dictate plant structure. An equipment enclosure may be used to provide a cleanroom environment supported by special air handling systems to keep the temperature, humidity, and air quality within desired limits. Thus, depending on the plant location or the type of product produced, the processing equipment may be contained in enclosures as illustrated in Figure 2-1.

Figure 2-1. Plant with Enclosed Construction

In contrast to these examples of enclosed plant construction, in many process industry segments, such as chemical processing and refining, it is common to locate the plant in a region characterized by a moderate climate and for some or all of the process equipment to be in the open, with no protection from the outside elements. In these cases the process equipment must be designed to operate without disruption in spite of changes in outside air temperature and occasional rain storms. The primary motivation for this type of open construction is the savings achieved by eliminating the housing for process equipment. When there is no need for the protection provide by an enclosure, the equipment will be located outside as illustrated in Figure 2-2.

Figure 2-2. Plant with Open Construction

When process equipment is located in the open, precautions must be taken to minimize heat loss. For example, the process lines that carry fluid and gas throughout the process are commonly encased by a heavy layer of insulation. To secure the insulation and provide added protection from physical damage, it is quite common to install a thin metallic covering, such as stainless steel, over the insulation. When viewing the plant from a distance, it may appear that the process lines are large stainless steel pipes but they are, in fact, much smaller pipes that have been insulated. Other process equipment such as reactors and heat exchangers that operate at elevated temperatures are also normally insulated to minimize heat loss and thus improve overall operating efficiency. Even storage tanks in the plant may be insulated if the material in a tank is normally above ambient temperature. In locations with open construction where the outside temperature may drop below freezing, it is often necessary to include electrical heating bands or small steam lines, known as steam tracing, around the process pipes and instrumentation sensing lines to ensure that the material in these pipes and sensing lines does not freeze.

Protection may also be required for the wiring associated with the field instrumentation that is used to measure the process conditions, such as pressure, temperature, flow and level, and with the field devices that are used to regulate those conditions throughout the process. The wiring to field devices for measurement and actuation is often supported and protected by cable trays. Electrical cables used to power motors in the process may also be run in cable trays. To avoid electrical interference, however, the cable trays containing instrument wiring normally do not contain electrical cables. Cable trays are typically distributed through the process in a manner that is designed to minimize the length of instrumentation wiring. Wiring from the cable tray to an individual field device is often enclosed in conduit to protect the wiring from physical damage as shown in Figure 2-3.

Figure 2-3. Cable Tray and Conduit for Instrumentation Wiring

The manner in which wiring is installed in the plant is dictated by electrical code. If there is a high probability of fire or explosion in the process because of the products that are produced, then the instrumentation wiring to field devices may be located underground to avoid wiring being damaged by fire or explosion. Such underground construction can make it difficult to add new field devices to the process unless spare cables are run during initial plant construction. Thus, there is some variation in the way the instrumentation and electrical power wiring is done depending on the industry and the standards followed in the country where the plant is located.

2.2   Plant Organization

When working in a process plant, it is important to understand and be aware of the physical division of the plant and the functional groups responsible for plant operation. In this section, an overview is provided of plant organization.

2.2.1   Process Areas

The basic unit of physical division within a continuous or batch plant is a functional grouping known as a process area. If a person is given directions on how to reach a piece of equipment in the plant, then these directions may include a reference to the process area where the equipment is located. For example, in a pulping plant for paper manufacture, the areas associated with bleaching or with log processing might be referred to as the "bleaching" or the "woodyard" areas of the plant. Area names are normally assigned during plant design to make it easier to identify different parts of the plant. In any discussion of plant operations, it is helpful to be aware of the names of the process areas within the plant.

> **Process Area** – Functional grouping of equipment within a plant.

In addition to giving each process area a name, it is customary to assign a number that also identifies the process area. For example, in a specialty chemical process plant, the carbonator area may be assigned Area number 177. In this case, this physical section of the plant would be referred to as the carbonator area or by its area number 177; the two references are interchangeable. As will be discussed in the chapter on plant documentation, many of the documents associated with control system design are organized by plant area. Also, the number assigned to a plant area is often contained within the identifier assigned to field devices and the area number and name are contained in the documentation for these field devices. The concept of process area divisions is fundamental to plant design and directly impacts the way the plant is described and documented.

2.2.2 Process Equipment

In some cases a process area is made up of multiple pieces of equipment that are similar in design and function. For example, the steam and electricity used in the process may be provided by multiple boilers and turbogenerators located in the powerhouse area. Nearly identical boilers may be used in the powerhouse to meet the changing steam and electricity demands of the plant. When multiple similar pieces of equipment are installed in a process area, these pieces of equipment are referred to as process units. In the powerhouse area, the boilers may be identified by unit number, for example, as power boilers 1, 2, and 3.

> **Process units** – Pieces of equipment in a process area that are similar in construction and function.

The number of process units that are contained in a process area may vary significantly depending on the product being manufactured and the plant capacity. In many applications, there are physical limitations that dictate the maximum capability of a process unit. Thus, to achieve a target production level, multiple units may be needed. Also, in some cases, such as power boilers, the best operating efficiency is achieved when the equipment is operating at its design steam production. So, by using multiple units as illustrated in Figure 2-4, it is possible to achieve better overall efficiency by varying the number of units in use rather than trying to use one larger piece of equipment that would be operating at low efficiency when the steam demand is less than normal.

Even though the process units making up a process area may appear identical in function and construction, it often turns out that there are differences, such as the positioning of instrumentation on the equipment, which impacts equipment performance and the commissioning of the control system. Even so, from the perspective of system design and process control, each process unit in a process area is treated the same.

**Figure 2-4. Multiple Power Boilers in
the Powerhouse Area**

In many processes, there's an uninterrupted flow of material through the process. As we saw in Chapter 1, this is referred to as a continuous process.

> **Continuous process** – Process that continuously receives
> raw materials and continuously processes them into an
> intermediate or final product.

This continuous flow of raw materials is typically converted into a product through heating, mixing, and reaction. A continuous process is designed to operate over a range of production and operating conditions. The production range is often quantified in terms of "equipment turn down."

> **Equipment turn down** – Ratio of the maximum to the minimum flow rate supported by a piece of equipment.

In many cases, the product produced using a continuous process may flow into a tank for storage before being packaged and shipped or before being further processed. In some cases, the output of a continuous process will be immediately used as the feedstock for a downstream continuous process. For example, the steam produced by a boiler is immediately used by other processes within the plant. Large storage tanks often separate process areas and provide a "cushion" that compensates for differences in processing rates.

As was also mentioned in Chapter 1, some products are produced using batch processes in which the final product is manufactured through a series of discrete steps. Fermentation processes, such as are used in the manufacture of beer, and other key processes in the food industry are often designed as batch processes. Batch processing is also used in many other industry segments, such as specialty chemicals and pharmaceuticals.

> **Batch process** – Process that receives materials and processes them into an intermediate or final product using a series of discrete steps.

In a batch process, the use of multiple process units may provide added flexibility in scheduling production. For example, in the specialty chemical industry the production schedule may be made up of many small batch runs in which different products are manufactured. Batch control systems that are designed to support the concept of process units commonly allow standard unit templates to be defined that represent the particular instrumentation used with each unit. Thus, the process unit plays an important role in batch recipe definition.

2.2.3   Plant Operator

Plants in the process industry represent a large capital investment. Achieving the desired return on investment in a plant depends heavily on consistently achieving the production targets that were the basis of the plant's design. The people who are ultimately responsible for minute-to-minute operations are the plant operators, who thus play a key role in a plant's operation and its commercial success. The operator's job is to ensure correct plant operation in one or more process areas under all processing conditions. Also, the operator is responsible for changing area production rates to achieve planned plant production targets. If production in a process area is not on target and/or equipment is damaged due to improper action being taken in plant operation, the operator responsible for this area of the plant stands a good chance of receiving a reprimand.

An important part of the plant operator's job is to constantly monitor process operation and to work through the control system to make the adjustments necessary to maintain correct operating conditions for safety and environmental compliance while attaining scheduled production levels.

> **Plant operator** – Person responsible for the minute-to-minute operation of one or more process areas within a plant.

The number of process areas assigned to an operator varies depending on process complexity, the number of pieces of equipment within each process area, and the degree of automation that is provided by the process control system.

The plant operator normally works from a control room that contains an interface to the process control system.

> **Control room** – Room in which the plant operator works.

Traditionally, control rooms have been physically located close to the process areas that they control. For example, if an operator is responsible for steam and electrical generation, then the control room might be located in the powerhouse area of the plant. In contrast to this, many large refineries have consolidated all their area control rooms into a single central control room. Other segments of the process industry are also trending this way.

In his or her job, the operator interacts with the control system to (for example) start and stop pumps, open and close valves and change the operating targets maintained by the control system. To successfully perform these functions, the operator must have a thorough understanding of the process area's operation and its physical layout. To achieve this level of understanding, an operator must have years of hands-on experience with the process, experience often gained by starting work in the plant in an entry-level position know as an operator helper and gradually working his or her way up to the position of plant operator. In addition, there may be different levels of plant operators, depending on the company. For example, in some plants there will be an Operator I, an Operator II, and an Operator III. To achieve certification for each operator position, it is often necessary to pass formal tests on the process and the associated control system. To advance, it is necessary to show proficiency at the next level of operation.

As an engineer, when you are working in and around the control room, it is important to always remember that the plant operator is ultimately responsible for running the plant. Thus, any changes in the process or control system by a control or process engineer or instrument technician must be done with the permission of the plant operator. In many cases, it is necessary to initiate a work order to formally request permission to make a change of any significance in the control system or the associated instrumentation. Depending on company policies, such work orders may need to be reviewed by a safety panel to ensure that the work can be done without introducing a safety risk into plant operation.

When installing or commissioning a control system, it is a good practice to work closely with the operator since the operator is ultimately responsible for the process area operation. A plant operator may have ten or twenty years' experience working with the process and understands things about the process that are not apparent from examining plant documentation. Where possible it is beneficial to directly involve the operator in any discussion of changes that are being made to the control system during installation or commissioning. Unfortunately, a person who is unfamiliar with plant operations may make the mistake of ignoring or bypassing the operator. For example, as a new engineer coming into the plant, you may discount an operator's input because the operator does not have an engineering degree. The authors have found that when you are working in a plant environment, it is always a good idea to have the operator on your side. It is important to talk to the operator as a peer, to ask him or her for information about areas you're unsure of about the process, and to learn from his or her knowledge and experience. By including the operator in these discussions and leveraging their understanding, there is both less risk of doing the wrong thing and better buy-in by the operator. In fact, by including the plant operators, they will go out of their way to enable changes that are needed in the control system and thus will contribute in a positive way to this work.

2.2.4 Supporting Department

Within the plant there are normally multiple departments that support the process operation and the instrumentation and control system. Depending on the size of the plant, there may be one or more maintenance shops that include instrumentation and control groups. Usually associated with a maintenance shop is a storage area that contains spare parts and spare field instruments. Also, the maintenance shop may include a machine shop that contains the tools and equipment needed to repair valves and other mechanical components critical to plant operation.

Most process plants include one or more laboratories that are centrally located in the plant or distributed throughout the process areas. Labs are critical to the plant operation since lab analysis of raw materials (feedstock) and process product may be required to identify variations that could impact the process operation. In addition, automated process analyzers may not be available to provide continuous, on-line measurements of the values associated with process product or feed material properties that must be controlled. Such on-line analyzers may be too expensive, may have proven to be unreliable, or simply may not be practical because of the complexity of the sampling system or the complexity of the analysis. As a result, manual sampling and lab analysis are often a requirement. When lab analysis is required to support plant operations, samples of the associated material are manually taken from the process and then transported to the lab for analysis. The collection of these samples, known as grab samples, is normally performed by the operator helper or a lab technician. Results from the lab analysis of these samples are used by the operator to adjust the process to meet quality requirements.

> **Grab sample** – Manually acquired sample of process product or feed material taken for lab analysis.

If the steps associated with an analysis are straightforward, then in some cases the grab sample may be processed by the operator helper at a test stand physically close to the process. In such cases, the analysis is often referred to as "near line" analysis.

One of the key objectives of a process control system is to maintain the product quality parameters within specifications. When a measurement of product quality is only available from the lab, the delay associated with processing the sample and the fact that this measurement is not continuously available impacts the way this information may be used to control the process.

In many cases, these lab measurements are used by the operator to manually make corrections in the process, such as changing feed flows or operating targets (called set points) such as temperature that impact the parameter(s) reflected in the lab test. However, specialized techniques do exist to automatically adjust the process based on entering the results of lab analysis into the control system. Where the capability is supported to automatically correct process operation based on lab results, an interface in the control system is normally provided that allows the operator or lab technician to manually enter lab results into the control system; the control system then makes the appropriate changes. Alternatively, when a Lab Management System exists, lab testing equipment may be interfaced directly with the control system to allow the results of the lab analysis to be automatically communicated to the control system as soon as the grab sample is processed. Some of these techniques for dealing with lab analysis are addressed in Chapter 4, On-line Estimator. An example of a lab for analysis of grab samples is shown in Figure 2-5.

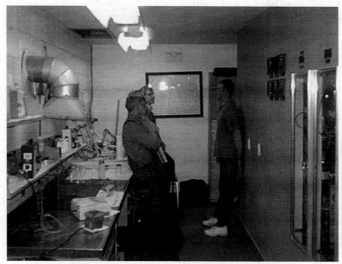

**Figure 2-5. Lab Used to Provide Analysis of
Quality-related Parameters**

2.2.5 Work Practices

When you are working in a plant, there are rules and procedures that
must be followed to ensure a safe work environment. Some general rules
apply to all plant situations. However, each plant has its own site-specific
rules that are often covered in a safety video that must be shown to anyone entering the plant for the first time. After the video is viewed, a test
may be administered on the material covered in the video. Only after
passing this test will a visitor be issued a pass that allows them to enter the
plant. It is also common for visitors to be allowed to work in the plant only
when they are accompanied by a plant employee who is familiar with the
plant layout and the safety regulations.

In many plants, there are restrictions on the type of clothing that may be
worn; for example, long sleeved shirts or safety shoes may be required. In
a facility where a hazardous gas release is possible in a section of the plant,
visitors working in that section may be issued a respirator and given
instruction in its use. In such cases, the plant may not allow a beard to be
worn since this could prevent the respirator from fitting snugly around
the face.

Most of the requirements associated with working in a plant are pretty
much common sense, and in addition, visitors can expect to be made
aware of plant rules and procedures. However, a new engineer who has
never worked in a plant environment may find the paper "A Guide for
Doing Onsite Work" [1] to be useful reading. This paper was written by
field engineers who have many years of experience in control system com-

missioning and startup. Many of the guidelines discussed in this paper are especially applicable to engineers doing this type of work.

2.3   Early Control Systems

Many of the techniques that form the basis of industrial process control have a long and interesting history. The original concept of feedback control that are addressed in later chapters is often attributed to work by Ktesibios in Greece around the third century BC when he implemented a float valve regulator to maintain a constant level in a vessel [2]. Even at this early time, there was a need to automatically maintain a process parameter (the level of fluid in a tank) at a desired value. In terms of today's control systems, the methods applied at that time, using mechanical linkages, would be known as proportional-only control as addressed in later chapters.

When steam power came about during the Industrial Revolution starting in the early 1700s and going though the early 1900s, there was a real need for automatic control. The generation and use of steam power first used in the early 1700s is a coordinated effort in which fuel is burnt with an appropriate amount of air, steam is generated, and the steam is then used to power a steam engine that drives mechanical devices within the plant. The amount of steam required is a function of the plant production rate or throughput. Thus, there was a need to adjust the amount of energy delivered to the process, that is, to be able to maintain various speeds of operation. The mechanical governors developed to automatically maintain operational speed under varying conditions of load are another example of proportional-only control implemented with mechanical devices. Such control was successfully applied in the regulation of rotative steam engines [3].

The modern process control system has its roots in developments that go back to the late 1800s and early 1900s. For example, Fisher Controls, which is one of the largest manufacturers of regulating control valves in the world, was initially established in the late 1800s by William Fisher. Mr. Fisher was a volunteer fireman in a small town in Iowa. Every time there was a fire, the water pressure to the town dropped as a result of the increased water demand. To compensate for this drop in pressure, it was necessary to manually adjust a supply valve. So Bill Fisher, being a resourceful person, found a way to automatically maintain the water pressure. The mechanism he built sensed the pressure and automatically adjusted the valve to keep the pressure in the desired range. From this humble beginning, he started building pressure regulators, valves, mechanical regulators and later pneumatic controllers. In the electrical power industry, Babcock & Wilcox developed techniques in the late 1800s for regulating the air used for combustion, thereby maintaining correct combustion conditions and reducing black smoke from smokestacks—an early example of proportional-only control that was initially implemented with mechanical devices.

The first large-scale automatic control systems used in the process industry were based on using air pressure for process measurement and valve actuation, and pneumatic logic circuits for control.

> **Pneumatic control system** – System that relies on a pressurized air supply for the operation of process measurement, control, and actuation devices.

Unlike mechanical control systems, pneumatic control systems allowed companies to place process controllers in control rooms located in the major process areas of the plant, remote from the measurement devices and actuators. A whole industry sprang up based upon developing and manufacturing controllers that were much more capable than the proportional-only mechanical control systems. The first versions of the proportional, integral, and directive-type (PID) controllers that will be discussed in detail in later chapters were first introduced at that time as pneumatic controllers. The ability to calculate parameters such as the average of two measurements was provided by computing elements based on pneumatic logic circuits. These devices could be used to implement many of the control techniques that form the core functionality of most distributed control systems today.

As electronic components became available for circuit design, some companies built electronic PID controllers using vacuum tubes. These early electronic controllers were costly, consumed a significant amount of power, and were not very reliable. Thus, these early electronic controllers found limited acceptance in the process industry. The first widespread use of electronic control systems came after the invention of the transistor and the commercialization of this technology. By the mid-'60s, electronics had advanced to the point where transistors and circuit boards could be created to duplicate what had long been done by pneumatic controllers and computing elements. The costs of these devices reached a level that made it practical to introduce self-contained electronic PID analog controllers and computing elements and, for the first time, analog electronic field devices.

> **Electronic control system** – System that relies on electric power for the operation of process measurement control and actuators.

With the introduction of electronic analog controllers, computing elements, and field devices, it was possible to eliminate the distance limitations imposed by pneumatic lines along with much of the installation and maintenance costs associated with a pneumatic control system. These electronic controllers were most often mounted in a long panel in the control room similar in many ways to the panels previously used to mount pneumatic controllers. The panel also included electronic paper chart recorders for critical process parameters, to allow any variations in operation to be permanently recorded. Pushbutton interfaces were mounted in the panel for the operator to start or stop pumps and open or close critical blocking valves and other functions, such as initiation of emergency shutdown of a unit, as illustrated in Figure 2-6.

By the early '70s, most of the new control system installations in the process industry were based on electronic analog control. Even so, pneumatic control can still be found even today in some stand-alone applications. Some pneumatic control systems in use today have been in service for 20 or 30 years and most have done a good job. However, in many cases these systems can no longer be maintained and are being replaced with electronic control systems.

Figure 2-6. Panel-based Analog Electronic Control System

Even in the early panel-based electronic systems, there was the concept of using alarms to notify the operator of an abnormal operating condition. Such notification was provided using annunciator panels mounted in the control panel. On detection of an alarm condition, the annunciator panel lit up, indicating the type of alarm (e.g., high pressure or high temperature). Panel-based electronic control systems were expensive to construct and required considerable floor space. Because of panel size and weight, it was very difficult to move or to make changes to the instruments and controls contained in the panel. To minimize the need for change, it was necessary to carefully design the control and alarm panel.

> **Control panel** – Wall containing controllers, annunciators, and other components used by an operator.

Early versions of chart recorders used circular paper on which the measurement values were plotted using ink pens to record selected process measurement values, as illustrated in Figure 2-7. As this technology developed, chart recorders were designed to use a long piece of folded chart paper on which multiple pens were used to record these measurements. The maintenance of both types of recorders was quite high and considerable time was required to change the pens and paper.

**Figure 2-7. Circular Chart Recorder
in a Control Panel**

In many cases, the control panels were quite long and were often filled with self contained PID controllers, recorders and indicator dials that the operator monitored by walking up and down the panel, looking to see if anything was wrong. Such an arrangement, and the physical length of a panel, often made it difficult for the operator to quickly respond to a process upset.

The control panel wiring was contained in the back of the panel. Because of space limitations, it was difficult to modify wiring or to troubleshoot the source of a problem in a controller mounted in the panel. An example of the wiring found in a panel installation is shown in Figure 2-8.

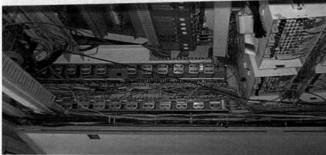

Figure 2-8. Control Panel Wiring

As illustrated in this example, the manner in which wiring was bundled together to save space could make it difficult to trace a wire. The labeling of wires and terminations was an essential part of panel design and installation. This labor intensive documenting of the wiring contributed to the cost of panel construction and associated system documentation.

2.4   Distributed Control Systems (DCS)

In the late '60s, computers were introduced into a few process plants for implementation of digital control. Since at that time computers were very expensive, the digital control tended to be centralized into one computer and was often restricted to a few critical control functions in continuous processes or to the automation of batch operations. Backup analog controls that supported local or computer control were used to avoid any loss of production when the computer was down because of a hardware or software failure. By the mid-'70s, electronics had evolved to the point where it was possible to integrate communication technology into the process control system and to distribute the control among multiple processors. Rather than accessing measurement values as an electric current, it became possible to digitally communicate measurement values and associated diagnostic information about the transmitter over a coax cable at fairly high baud rates. The enabling technology was the advent of reliable,

low-cost microprocessors that could be programmed to provide functionality previously achieved using many analog components and circuits.
The primary means of control, the PID controller, could be implemented
in software. The electronic devices known as multi-loop controllers were
introduced to address multiple measurement and control requirements
and thus eliminated the need for individual dedicated hardware controllers for each PID control function.

> **Multi-loop controller** – Device for interfacing to multiple
> field measurements and actuators for the purpose of calcu-
> lation and control.

The functions previously provided by dedicated hardware computing elements were now performed by software. Many of the leading control system suppliers to the process industry introduced distributed control
systems (DCS) in the mid- to late '70s.

> **Distributed control system** – Control system made up of
> components interconnected through digital communica-
> tion networks.

The expanded communications capability of these systems gave the system designer new freedom to distribute the controllers to remote control
centers located near to the process and thus reduce the length of wiring
runs to field devices. As a result of this distribution of the control system,
the size of control rooms and the associated construction costs were
reduced. In most cases, the control room only needed to be large enough
to contain the monitors and keyboards associated with the operator's
interface to the control system. The need for dedicated pushbuttons and
status indicators for motor control were eliminated since these types of
discrete measurement and outputs were integrated into the control system. It became possible for the operator to view and access all aspects of
the process and the control system from the control system monitors and
keyboard. In many cases, the only remnant of the control panel that
remained was hardwired buttons to shut down the process if operator
access to the control system was lost.

The introduction of the digital distributed control system had a major
impact on the tools that were available to the operator in performing his or
her job. The monitors and keyboard interfaces that were provided in a distributed control system became the window into the process and the only
means of working with the control system. Thus, as part of a distributed
control system installation, it was necessary to train the operator on this
new interface to the process. The functions provided by the panel-based
system were still provided but were implemented in a different manner.

Essentially the screens provided on the monitor replaced the information previously displayed through gauges, dedicated controller interfaces, paper chart recorders, and annunciator panels. This was a big change for the operators, that is, seeing the state of the process by paging through control displays instead of by looking along the panel.

To ease operator transition from panel-based systems to monitor-based systems, the manner in which information was displayed on the monitor mimicked the way this information had earlier appeared on the control panel. The control faceplate that is used in distributed control systems can be traced to the manner in which information was displayed in panel-based analog controllers.

> **Control faceplate** – Display screen that mimics the manner
> in which information was provided on an analog
> controller.

Details of a faceplate, such as showing the low-to-high engineering unit range of the measurement and displaying the value as a small vertical bar, replicated the presentation of the old analog controllers. Since the control interface provided by the control panel was familiar and had proven to be effective, a similar presentation was adopted in the screen design and display elements used by early distributed control systems. The annunciator panel was replaced by an alarm banner provided in the monitor. Even the circular dial and trend displays of historic information that are supported in most control systems today simply mimic some of the interfaces provided in the panel-based control systems.

When a distributed control system was installed in a plant that had previously had a panel-based control system, it soon became common to take advantage of the flexibility of the new system to improve and streamline plant operations. Control rooms were often centralized as part of the distributed control system. For example, in a plant that previously had four or five control rooms, each staffed with operators, it became possible to replace all of these rooms with one control room. As a result, the number of operators needed to run the plant could be significantly reduced. The scope of the operator's responsibilities was often increased to include more process units or process areas. This reduction in manpower was a cost saving for the plant, but more important, this centralization allowed better coordination of production between process areas and thus resulted in improved operating efficiency. Many companies in the process industry quickly adopted this technology for new installations and, where possible, upgraded existing plants with distributed control systems. Plants that did not adopt this technology are today having trouble competing with the ones that did because of the cost of having multiple operators and multiple control rooms and because they cannot take advantage of the improvements in operating efficiency. Over time, the maintenance associated with older panel-based control systems in such plants can also become an operational problem and added expense.

Figure 2-9 is an overview of the major components that make up a distributed control system.

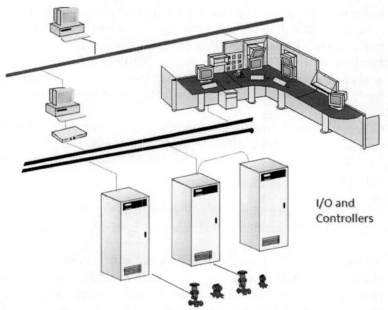

Figure 2-9. Distributed Control System Overview

The installation of a distributed control system changed nearly every aspect of the design, implementation, checkout, commissioning, operation, and maintenance of the control system, including the tools and skills that were required. For example, the volt-ohm meter was the primary tool that an instrumentation technician used in troubleshooting a panel-based system. When a fault was found, the physical PID controller or the computing element used with the controller could simply be replaced. Since a distributed control system depends heavily on the use of digital communications and functions performed in software by microprocessors, different skills were required to set up, operate, and maintain the control system.

When a distributed control system was first installed in a plant, there was often disagreement about who was going to work with the system software to define and maintain the functionality associated with measurement and control. The person who was previously responsible for maintaining the analog electronic control system, who knew how to use a volt-ohm meter, saw his job going away. So he said, "Well, I should be responsible for the distributed control system." But to set up and maintain a distributed control system required a different skill set. To work with a distributed control system, a person had to know how to use the tools provided with the system to set up the software to perform the measurement, calculation, control, and display. The significant impact this had can be illustrated by considering the operator interface design. When installing a panel-based control system, it was necessary to consider the instruments to be included in the panel, come up with a panel layout that fit the allotted space and meets operator interface requirements, construct the panel, and install the instruments. In contrast, when installing a distributed control system, the person responsible for the operator interface had to be familiar with the display capability of the control system and then design and construct displays that allowed the operator to access process information and interact with the control system in the most efficient manner.

In most cases, distributed control systems come with tools that allowed this setup to be done through a process of configuration, in which functionality may be defined and characterized without having to write software.

> **System configuration** – Specification of measurement, calculation, control, and display functions of a control system.

To learn the skills needed to do system configuration often meant going to classes provided by the control system manufacturer. In some instances, new control groups with such skills were established in the plant to perform control system configuration.

The transition to distributed control systems was equally disruptive for the plant operator. When working at a control panel, he or she would make adjustments in the process by pushing buttons or turning knobs to change operating points. Motors were started and stopped using pushbuttons on the panel. However, with the installation of a distributed control system, the operator interface was one or more monitors. To make a process change, he now had to type in the change using a standard or custom keyboard. A common reaction when an operator was being trained on how to use a keyboard was "That's not my job." In some cases the operator was unable to make this transition and was moved to another position in the plant.

To ease the transition, it was important to do extensive operator training as part of the control system installation. Special operator training simulators were often constructed from spare parts to allow the operators to work with the system interface in the training class. In some cases, these training simulators included a simulation of the process to provide a more realistic training environment.

One of the things that control system manufacturers and companies learned in transitioning to distributed control systems was that when a control system is installed in a plant, it is important to consider and plan for the changes introduced in the way people in the plant will perform their job function. The manner in which this transition is addressed can impact the acceptance of the control system and the benefits derived from the installation.

2.5   Operator Interface

With the introduction of the distributed control system, operator interface design was an important concern for the both the control system manufacturer and the company purchasing the system. The earlier systems only supported the capability of the operator interface to display faceplates and buttons that looked similar to those found on a control panel. From a design perspective, many of the same skills previously used in panel layout could be applied in deciding on the grouping of faceplates and buttons that were contained in a display.

When a company installed distributed control systems in multiple areas, it was quite common for a single operator to then be responsible for hundreds of loops, motors, and measurements. In such installations, the approach of only providing faceplates in the operator displays was found to be ineffective. The primary reason for this was that the only differentiation between the numerous displays in the system was the instrument description contained in the faceplate. For an operator to understand the significance of a measurement or control faceplate shown in a display, it was necessary for the operator to read this information. So, to address this problem, manufacturers soon introduced the concept of graphical displays. In a graphical display, it was possible to show a pictorial representation of the equipment and pipes used in the process and to show measurement values at the point where the measurement was made in the process. With a graphical display, the operator could immediately recognize the area shown and could quickly find the information within the control system.

> **Graphical display** – Display containing a pictorial representation of process equipment and piping along with measurement values at the point they are made.

Graphical displays were initially used in some of the early installations only to show information about the process. Faceplate displays were still used in these systems to show control information. The reason that faceplates were retained as the interface for control was due to the fact that faceplates often contain multiple pieces of information needed by the operator in making a change in the process. For example, the faceplate would typically contain the following information:

- Control parameter measurement value
- Control parameter target value
- Output of the control to an actuator
- Control measurement alarms
- Mode of control
- Status (Good/Bad Indication) of the measurement used in control

Over time, manufacturers and plants working with distributed control systems found techniques to allow all this information to be effectively included in graphical displays. To avoid clutter in the display, the status of measurement alarms was indicated by dynamically changing the color of process elements shown in the display. Rather than showing the mode of control, the indication of mode would only appear in the display if the actual mode was different from the normal or expected mode of operation. Using these techniques, it was possible to show control as well as measurement information on the same screen. In most recent installations of distributed control systems, the operator interface consists primarily of graphical displays, as shown in Figure 2-10.

Figure 2-10. Operator Interface Graphical Displays

When the operator is making adjustments in the operating targets maintained by the control system, he or she may need more information than is shown in a graphical display. For this reason, most manufacturers of modern distributed control systems supply predefined software components known as control dynamos as part of the tools provided for creating operator displays. These dynamos may be added to graphical displays to show some information about a control loop. When the operator selects a dynamo in a graphical display, a faceplate appears in the display area and provides further information about the control loop. Components of this dynamo, such as slew bars, buttons, and entry fields may be used by the operator to change control operating targets as illustrated in Figure 2-11.

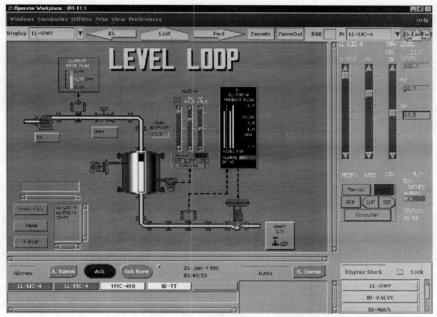

Figure 2-11. Control Dynamo Support in Graphical Displays

From the faceplate called up from a graphical display, the operator may be able to acquire additional information such as the alarm limits or the settings associated with a control. The manner in which such graphic components are defined and used varies with the distributed control system manufacturer.

The distributed control system, introduced in the '70s and '80s, has reduced the overall cost of control system, design, installation, operation and maintenance. As these systems have evolved the cost of purchasing and maintaining control systems has continued to be reduced.

In early distributed control systems, the electronics, keyboards, and furniture to house these components were custom designed as illustrated in Figure 2-12.

Figure 2-12. Custom Furniture Used in a Control Room

The custom components used in the operator interface were often expensive to install and maintain. As personal computers became more common and the processing speed of these devices increased, most manufacturers of distributed control systems migrated toward the use of standard components such as personal computers, commercial operating systems, standard keyboards, and standard monitors for the operator interface. Only for some special applications can the cost of customizations still be justified. In most plants, the standard interface is the regular keyboard. This trend has been reinforced by the fact that personal computers are widely used at home and so today, the workforce is already familiar with computers and the mouse.

2.6   System Installation

As we have seen, with the introduction of distributed control systems, it became possible for the multi-loop controllers that interfaced with the field devices to be physically located away from the control room. These multi-loop controllers were mounted in cabinets and thus appeared as racks of equipment. The racks were housed in rooms that were built to protect the multi-loop controllers from the plant environment and were located near the field devices. These became know as rack rooms.

> **Rack room** – Room designed to house control system components.

In some cases, the motor control centers that were located throughout the plant to house electrical equipment, such as motor starters, also served as rack rooms. In such cases, it was important to take sufficient precautions to avoid electrical interference. One of the concerns of mounting control equipment close to a motor control center is electromagnetic energy that's given off when the breakers and contactors switch.

To keep the multi-loop controller electronics within their designed operating temperature range, the rack rooms normally came equipped with air conditioners and heaters. In some process areas, it was also necessary to provide chemical air scrubbers in the air handling system. For example, in pulp and paper plants the concentration of Sulfur dioxide (SO_2), Total Reduced Sulfur (TRS), chlorine (Cl), and other components found in the air around a recovery boiler or bleach plant can damage circuit boards and contacts. An example of a rack room is shown in Figure 2-13.

**Figure 2-13. Rack Room for
Distributed Control Equipment**

As was true in the early days of distributed control systems, today the long-term reliability of a distributed control system can be impacted by the effectiveness of the precautions taken to protect the electronics from electrical interference, extreme temperatures, and corrosive air components. In selecting a rack room site and requirements for the room, the design should take these considerations into account.

The layout and design of the multi-loop controllers used in early distributed control systems varied depending on the equipment manufacturer. In general, these controllers consisted of the following components:

- **Controller boards** – Contained the memory and processor used to perform measurement, calculation, and control functions.

- **I/O (input/output) boards** – Contained electronics to convert the field measurements into a digital value that could be used by the controller and to convert digital outputs to electric signals provided to field devices.

- **Terminations** – Wiring junctions for field wiring and signal processing to provide an interface to I/O boards.

An example of the terminations provided on these earlier systems is shown in Figure 2-14.

Many systems were designed to support redundant controller and I/O boards. Redundancy improved overall system reliability in that the failure of a single board would result in a backup board automatically assuming the function of the defective board.

Figure 2-14. Controller Termination Boards

The measurements made by field devices could be brought into a distributed control system in several different ways. In some cases, the measurement was represented as a voltage signal. For example, a thermocouple measurement element provides a low-level signal measured in millivolts. More commonly, the field measurement was provided to the analog input modules of the control system as a 4–20 mA current signal. Later systems included support for digital communication with the field devices. For example, some of the digital field devices introduced in the late '80s support digital communications using the HART (Highway Addressable Remote Transducer) protocol. A variety of discrete measurements were also supported by the early distributed control systems for interfacing to motor starters, valve limit switches and other field devices that utilize discrete (on-off) input and output signals. Table 2-1 is an example of the types of field inputs and outputs supported by the early distributed control systems.

Table 2-1. Typical I/O Supported by Early DCS

Analog Input Modules	Discrete Input Modules
• High-Level Single-Ended	• 3 - 32 Vdc w/ debounce
– 1-5 Vdc	• 3 - 32 Vdc w/o debounce (fast switching)
– 4-20 mA	• 3 - 32 Vdc for Vortex flowmeters
• High-Level Isolated	• 90 - 140 Vac/Vdc
– 0-10 Vdc	• 180 - 280 Vac/Vdc
– 4-20 mA	• Dry Contact input (low side switching)
• High-Level Isolated	
– -10 to 70 mV	**Discrete Output Modules**
• Various RTD types	• 24 Vdc
• Various Thermocouple types	• 3-60 Vdc
	• 24-140 Vac
Analog Output Modules	• 24-140 Vac w/ MOV-protection
• 4 - 20 mA dc	• 24-280 Vac
	• 24-280 Vac w/ MOV-protection
Smart Modules	• Relay Output
• HART Input	
• HART Output	

Wiring from the field devices to the control system varied, depending on the industry and the engineering firm directing the installation. Where the air in the plant might contain corrosive elements, the company could require that a continuous wiring run be used for all connections between field device and the I/O termination at the control system. However, it was (and is today) much more common for the field wiring to be terminated in a junction box located in the rack room or near the process and for separate wiring to be used for connection to the control system. Providing an intermediate junction box allowed wiring to the control system to be changed without having to change the wiring to the field devices. An example of a junction box for field wiring is shown in Figure 2-15.

Field junction boxes may be found in plants that cover a large area, such as refining and chemical plants. In installations that use junction boxes, the wiring from the junction boxes to the control system may be done using multi-conductor cable to reduce installation cost.

Figure 2-15. Junction Box for Field Wiring

2.7   External System Interfacing

As was true in the early days of distributed control systems and continues to be true today, the measurement, calculation, and control capabilities of a distributed control system may be used to address the needs of a wide variety of applications in the process industry. Even so, some pieces of equipment such as compressors or chillers come equipped with a dedicated control system. In addition, there are some functions associated with high-speed discrete processing that are often better addressed by integrating specialized external systems into the distributed control system. For example, the high-speed requirements associated with packaging operations, or mechanical processes such as the transfer of material in the woodyard of a pulp and paper plant, may require high-speed discrete control supported by a programmable logic controller (PLC) as illustrated in Figure 2-16.

Some of the very first distributed control systems integrated PLCs by using controller discrete inputs and outputs wired to PLC discrete outputs and inputs. However, in the early '80s serial interface cards were developed by DCS manufacturers that allowed blocks of registers and individual outputs in the PLC to be quickly read and written using a single serial interface cable. For the companies that introduced this capability, it pro-

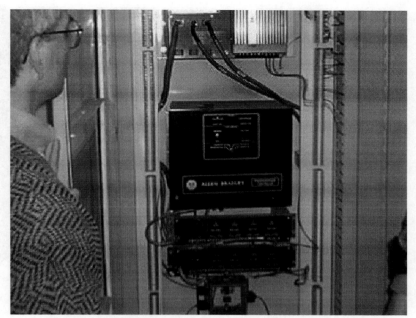

Figure 2-16. Addressing Specialized Applications with a PLC

vided a significant technical advantage in device integration. These interfaces were also often used to tightly integrate motor control and safety systems that were based on PLCs.

In a process plant, there are often specialized stand-alone equipment and systems, such as vibration monitoring and shutdown systems. Such a system is shown in Figure 2-17. Today, plants routinely integrate these types of equipment and systems into the DCS using the serial interface cards supported by multi-loop controllers. The advantage of integrating this type of equipment information into the DCS is that measurements, information on detected faults, the status of interlocks inputs, and so on, can be included in the same displays that the operator uses to monitor and control this piece of equipment. Through such integration, it is often possible to eliminate the need for a dedicated control room monitor for this specialized equipment. This saves space in the control room. More importantly, information for the operator can be organized in a consistent way for viewing. As a result, in many cases the operator is able to respond more quickly to abnormal situations and to better use information provided by stand-alone equipment and systems.

It may not be economically feasible to automate all aspects of a manufacturing process. For example, in the batch industry, it may be necessary at some point for an operator helper to manually add a bucket of material to the batch. In a powerhouse, it may be necessary for the

Figure 2-17. Vibration Monitoring System

helper to perform manual steps on the fuel system before a boiler may be started up. In these instances, an interface may be located in the process area to allow these local actions to be coordinated with the control room operator. Through these local interfaces, it may be possible for the operator helper to access information in the DCS and also for the control room operator to see actions taken by the operator helper. For example, once a manual operation is complete, the helper may use the local interface to acknowledge that he has completed it. An example of a local interface is shown in Figure 2-18.

To facilitate coordination between the control room and the control of a field device, such as a pump or conveyor motor in a continuous process system, the control within the DCS may be implemented using a "Hand/ Off/Auto" switch (HOA). If the control room operator selects the "Hand" position, he or she gives permission for local starts and stops. The "Auto" selection is used to indicate that the control system may start and stop the device. "Off" is selected by the operator to turn the device off under all conditions. In batch process systems, similar mechanisms are used to coordinate local actions with actions taken by the control system.

2.8 Modern Control Systems

By the mid-'90s, the expanded use of Ethernet in industry and improvements in price and performance made it practical to use this technology to replace proprietary communication systems incorporated in the first generation of distributed control systems. Also, new operating systems for personal computers became available that could be used to

Figure 2-18. Local Interface

satisfy the requirements associated with the operator interfaces and stations from which the distributed control system is configured. In addition, advances in microprocessors, surface-mounted integrated circuits, and electronics packaging allowed the boards and components used for multi-loop controllers to be greatly reduced in size. At the same time the increased performance of the microprocessors used in the controllers allowed more calculations to be done within a multi-loop controller. Based on these improvements, a whole new generation of distributed control systems was introduced in the mid-1990s to the process industry. As illustrated in Figure 2-19, these changes in technology allowed the same capability to be housed in a much smaller space. Thus, by going to this technology, it was possible to further reduce the installation cost of a distributed control system.

The packaging of the latest generation of multi-loop controllers supports much denser termination than was possible in earlier distributed control systems. Also, the terminations in these later systems are designed to allow an I/O card to be installed and removed without disrupting the field wiring or the operation of the controller. An example of the terminations and wiring to the modern multi-loop controller is shown in Figure 2-20.

Figure 2-19. Comparison of Multi-loop Controller Designs

Figure 2-20. Field Wiring and Terminations

In the same timeframe that process control manufacturers were working on a new generation of distributed control systems, the manufacturers of field devices were working to provide a new generation of devices based on a variety of fieldbus technologies, such as Foundation Fieldbus, Profibus, Asibus, and DeviceNet. These technologies allowed access to device information using high-speed communication. Also, an important feature of these bus technologies was the fact that multiple devices could be accessed using a single pair of wires. Control system manufacturers working on this new generation of DCS incorporated full support for these new technologies into their multi-loop controller and I/O card designs. As devices based on fieldbus technology emerged, it became possible to further reduce the field wiring to multi-loop controllers, as illustrated in Figure 2-21.

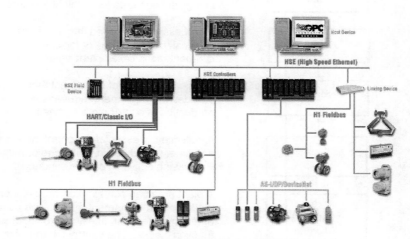

Figure 2-21. Field Wiring Using Bus Technology

2.9   The Impact of Standards

Many of the control standards developed in the mid-'90s had a large impact on the design and implementation of distributed control systems. For example, in the batch processing industry, the ISA-88 series of standards defined terminology that provided a consistent way to design and document batch control systems. This was especially important to international companies that manufacture in multiple locations. As previously mentioned, the concept of a process area is common to both batch and continuous processes. However, to address batch processing requirements, ISA-88 extended this concept to process areas that can be further divided into process cells, units, and equipment. The identification of cells in continuous plants isn't all that common, but in batch plants, it is. Another important construct defined by ISA-88 was the "module." In both continuous and batch control systems, the module acts as a container of measurement, calculation, and control implemented as function blocks.

> **Control module** – Container of measurement, calculations, and control implemented as function blocks. A module may contain other control modules.

An overview of the terminology defined by ISA-88 is shown in Figure 2-22.

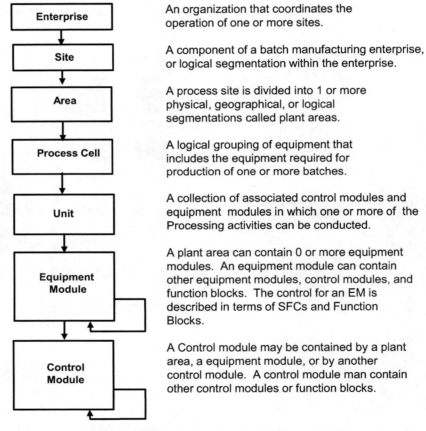

Figure 2-22. Overview of ISA-88 Terminology

Many companies and control system manufacturers have adopted the terminology defined by ISA-88 for control system design and implementation. An excellent reference on ISA-88 has been published by ISA and may be read to learn more about this series of standards. [4, 5, 6, 7]

The IEC 61131 (International Electrotechnical Commission) standard released in the early '90s had a significant impact on the design of the latest generation of distributed control systems. The IEC 61131 standard defined four languages that can be used to define control. A graphical technique based on the use of function blocks was defined for the design and implementation of measurement, calculation, and control.

> **Function block** – A logical processing unit of software consisting of one or more input and output parameters.

One of the graphical control language defined by IEC 61131 is sequential function charts (SFC) for the implementation of sequential operation in batch or discrete control logic. The use of sequential function charts was adopted by ISA-88 for use in the definition of batch control logic. The IEC 61131 standard also defined ladder logic, which is mainly used in PLCs, and structured language, which is similar to a programming language and can be used to define calculations.

The function block language defined by IEC 61131 was quickly adopted by many process control system manufacturers as the primary means of configuring measurement, calculation and control in a distributed control system. The sequential function chart language was included in many systems, primarily for the implementation of batch control. The ISA-88 series of standards also adopted sequential function charts as a tool for defining batch control logic.

The IEC 61131 work on function blocks was further refined in the ANSI/ISA-61804 standard that defines function blocks for use in the process industry [8]. The work on the ANSI/ISA-61804 standard was the basis for the Fieldbus Foundation's specification for the function block application process. As a result, the Fieldbus function block specification is fully compliant with the IEC 61804 standard. The IEC 61131 and ANSI/ISA-61804 standards provide a graphical means for defining measurement, calculation, and control as illustrated in Figure 2-23.

The Fieldbus Foundation Function Block Application Process specification is written primarily for developers of field devices. Part One of this specification defines the architecture or framework for function blocks. It includes an object-oriented model (i.e., a design based on the use of data structures consisting of data fields and methods) of the key components that are found in all function blocks. Part Two defines the function blocks that are most often used for measurement and control applications. Part Three of the specification is dedicated to less frequently used function blocks that may be required for more advanced calculation and control applications.

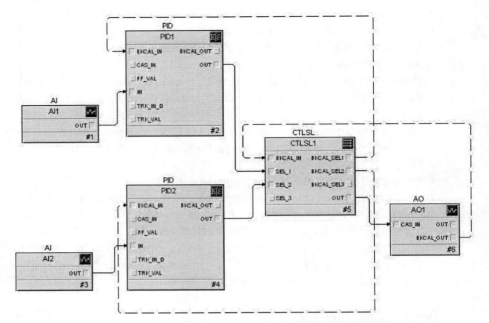

Figure 2-23. Function Blocks in Control Applications

The Fieldbus Foundation specification was written by engineers that represent the major manufacturers of field devices and distributed control systems. For example, the initial development teams that worked on Parts One and Two of the specification included engineers from Siemens in Germany, Yokogawa in Japan, and Emerson in the United States. Other companies such as Invensys and Honeywell contributed to Part Three. So, the blocks defined by these specifications represent the experience of many individuals from the major companies in the process control industry. The function blocks defined by this specification are supported by the major suppliers of distributed control systems.

The Fieldbus Foundation function block set is used in some distributed control systems as the primary tool for the creation of modules. The benefit of this approach is that measurement, calculation, and control may be defined independent of whether the associated function blocks are assigned to execute in Foundation fieldbus devices or in a multi-loop controller. The engineering and maintenance environment for working with function blocks allows the user to graphically represent measurement, calculation, and control as one or more blocks connected together, similar to the concept of construction using LEGO® blocks. Information flow between blocks is shown as wire connections between input and output parameters displayed on blocks, with inputs shown on the left side of the block and outputs shown on the right side of the block. The flow of information from one function to the next is a very natural way to structure measurement, calculations, and control.

The block set defined in Part Two of the Fieldbus Foundation function block specification may be used to address the most common application requirements. The manner in which these blocks may be applied to address common control applications will be discussed in a later chapter. The function blocks defined in Part Two of the Fieldbus Foundation function block specification is shown in Table 2-2.

Table 2-2. Basic Function Blocks

Discrete Input	P, PD Controller
Discrete Output	Control Selector
Analog Input	Manual Loader
Analog Output	Bias/Gain Station
PID, PI, I Controller	Ratio Station

As has been mentioned, the blocks defined in Part Three of the Fieldbus Foundation function block specification may be used to address more advanced calculation and control applications. In later chapters, we will discuss the blocks that may be used in advanced control, and show application examples that illustrate how these blocks have been used to meet control requirements. The blocks defined by Part Three are listed in Table 2-3.

Table 2-3. Advanced Function Blocks

Pulse Input	Deadtime
Complex Analog Output	Arithmetic
Complex Discrete Output	Calculate
Step Output PID	Integrator (Totalizer)
Device Control	Timer
Setpoint Ramp Generator	Analog Alarm
Splitter	Discrete Alarm
Input Selector	Analog Human Interface
Signal Characterizer	Discrete Human Interface
Lead Lag	

References

1.   Gibbs, J., Thorp, S., and Keels, B. "A Guide for Doing Onsite Work," *Chemical Engineering*, Feb. 1990.

2.   Bissell, C. C. "A History of Automatic Control," *IEEE Control Systems*, Pages 71–78, Apr. 1996.

3.   Wheeler, Lynder Phelps (1947), "The Gibbs Governor for Steam Engines", in Wheeler, Lynder Phelps; Waters, Everett Oyler; Dudley, Samuel William, *The Early Work of Willard Gibbs in Applied Mechanics*, New York: Henry Schuman, pp. 63–78.

4.   Hawkins, W. M. and Fisher, T. *Batch Control Systems – Design, Application, and Implementation*, 2nd Edition, Research Triangle Park: ISA, 2006 (ISBN: 978-1-55617-967-9).

5.   ANSI/ISA-88.00.01-2010, *Batch Control Part 1: Models and Terminology*.

6.   ANSI/ISA-88.00.02-2001, *Batch Control Part 2: Data Structures and Guidelines for Languages*.

7.   ANSI/ISA-88.00.03-2003, *Batch Control Part 3: General and Site Recipe Models and Representation*.

8.   IEC 61131, *Programmable Controllers Package*.

9.   ANSI/ISA-61804-3 (104.00.01)-2007, *Function Blocks (FB) for Process Control – Part 3: Electronic Device Description Language*.

10.   ANSI/ISA-TR61804-4 (104.00.02)-2007, *Function Blocks (FB) for Process Control – Part 4: EDD Interoperability Guideline*.

3
Measurements

In this chapter, the focus is on a few of the field devices that are used to provide the process control system with measurements of flow rate, pressure, temperature, and level. The objective of this chapter is to give the new engineer some sense of the types of field devices used to measure process conditions in a plant environment.

The field devices used to measure process operating conditions are commonly referred to as transmitters.

> **Transmitter** – Field device consisting of a housing, electronics, and sensing element that is used to measure process operating conditions and to communicate these values to the process control system.

The performance that may be achieved by a control system may be impacted by the measurement limitations of the transmitters selected to provide the process measurements. For example, to achieve target product specifications, it may be necessary to maintain the temperature of a process stream to within 0.1 degrees centigrade. However, if the transmitter (which includes the sensing element or sensor) selected for the measurement is not designed to provide a temperature measurement value with this level of accuracy or repeatability, then it may not be possible to achieve the target product specifications.

> **Accuracy** - The ability of a measurement to match the actual value being measured.

> **Repeatability** - The closeness of agreement among a number of consecutive measurements of outputs for the same input.

During the design of a plant, the design engineers select transmitters and actuators that match the process requirements. They typically have many years of experience working with different applications. Based on this field experience and manufacturer recommendations, they make the decision as to which measurement technology best satisfies the application measurement requirements. This decision is typically given a lot of thought, with the objective being to provide easily maintained measurements with the necessary level of accuracy and repeatability for the expected operating conditions.

The limitations imposed by the transmitters will become more apparent in later chapters that focus on control. In particular, the resolution and accuracy of the transmitters used for measurement are shown to directly affect the quality of control. Transmitter selection during plant design is important since these decisions have a long-term impact on plant operation, independent of the control system or the control technique that is selected for process control. In this chapter, the limitations of each measurement technology will be examined from the standpoint of how these impact control performance.

3.1   Magnetic Flowmeter

One of the most fundamental process measurements to the process industry is the volumetric flow rate of a liquid stream. For some applications, the volumetric flow rate of a liquid stream can be measured using a magnetic flowmeter. One of the attractive features of a magnetic flowmeter is that the liquid flows through the meter without any restriction in the pipe (which is not the case with some flow measurement devices). Therefore, there is very little, if any, pressure drop associated with the measurement. This is important because restrictions make it more difficult for the pumps to keep the liquid flowing through the process. Also, to avoid flashing in the process line, it may be important to minimize pressure drop if the liquid temperature is near its boiling point at the upstream line pressure.

> **Flashing** - Change from a high pressure liquid to a low pressure liquid/gas mixture.

Where the liquid properties allow a magnetic flowmeter to be used, it is often the preferred measurement device when an accurate flow measurement is needed over a wide operating range. An example of a magnetic flowmeter is shown in Figure 3-1.

Figure 3-1. Magnetic Flowmeter

As the name implies, the principle that is used by a magnetic flowmeter to sense volumetric flow rate is based on the movement of a conductive material through a magnetic field. Coils of wire embedded in the outer shell of a magnetic flowmeter are used to induce a very strong magnetic field in the liquid stream that flows through a pipe that forms the center of the meter. In an electrical generator, a voltage is induced in a wire as it passes through a magnetic field. Similarly, when a conductive liquid flows through a strong magnetic field, a voltage is induced in the liquid stream that may be measured using a small electrode mounted flush with the wall of the flowmeter. The flow rate of the material is determined based on the induced voltage and the known strength of the magnetic field.

Liquid streams that are characterized by low electrical conductivity, such as pure water, are not good candidates for magnetic flow measurement. This requirement that the liquid be conductive limits the applications for which it is appropriate to specify magnetic flowmeters.

A local source of power is required for a magnetic flowmeter, that is, these transmitters are typically designed as "four-wire" devices.

> **Four-wire device** – A field device that requires a local source of power in addition to the pair of wires used for communication with the control system.

The term *four-wire device* implies that two wires are used to communicate the measurement to the control system digitally or as a 4–20 mA signal that is proportional to the measurement value. In addition, a second set of wires is used to power the magnetic flowmeter, usually through an AC power line. Considerable power is required to produce the magnetic field used by the magnetic flowmeter.

The manufacturers of magnetic flowmeters provide application guidelines that detail the installation requirements. This information includes limitations on the liquid stream composition and on operating conditions, such as the operating pressure range and the temperature of the liquid stream. When using a magnetic flowmeter, the liquid pipe line must be kept full under all operating conditions. Air bubbles in the liquid stream may cause significant spikes in the indicated flow measurement, that is, the flow rate may be constant but air bubbles may cause a false indication of a high flow rate. If a false indication of flow rate is used by the control system, then inappropriate action may be taken that could disrupt the process. Attention to detail in device installation is necessary to help ensure a reliable measurement.

The location where a magnetic flowmeter is to be installed in the pipe is normally fitted with flanges. This allows the meter to be placed between the pipe flanges and then bolted into place. To avoid turbulent flow, which could also potentially disrupt the flow measurement, there is typically a requirement for a minimum distance of straight pipe upstream of the flow measurement. This minimum distance of straight pipe may be expressed as a multiple of the pipe diameter and is normally indicated in the installation instructions provided with the device.

3.2   Vortex Flowmeter

The vortex flowmeter may be used to measure a liquid or gas stream. This measurement is based on the phenomenon that vortices will form around a restriction placed in a liquid or gas flow stream. An oscillation is established by the gas or liquid stream vortices that are detected by the measurement element. The measured volumetric flow rate is based on oscillation frequency, stream composition, and operating temperature and pressure. A great deal of engineering and technology go into the design of a vortex flowmeter to ensure a reliable measurement, because these types of transmitters may be installed in process lines that are vibrating due to pumps or the operation of other mechanical devices in the process.

The pressure and temperature rating of a vortex flowmeter may depend on the device manufacturer. For example, common operating conditions for liquid flow measurement may be addressed by a large variety of manufacturers. However, only a limited number of manufacturers may offer devices that are designed for high pressure and temperature applications such as steam flow measurement. An example of a vortex flowmeter is shown in Figure 3-2.

The vortex flowmeter may be used where there is a need to measure flow over a wide operating range. Manufacturers of devices used for flow mea-

Figure 3-2. Vortex Flowmeter

surement often express the expected range of operation in terms of turndown. For example, the product data sheet for a vortex flowmeter may
specify that the device turndown is 38 to 1. In connection with flow measurement, the term *turndown* is defined as the ratio of the maximum to the
minimum flow rate that may be reliably measured by the device.

> **Turndown** – Ratio of the maximum to the minimum flow
> rate that may be measured by a transmitter.

In a discussion of flow measurement, someone may say, "Well, this device
has pretty poor turndown. It has a 5-to-1 turndown." In this example, a
device turndown of 5 to 1 means that it cannot reliably measure flow rates
below 20% of the maximum flow rate. If the flow in a process will vary
over a wider range then it becomes important to use a device such as a
vortex flowmeter that is characterized by a large turndown. For example,
if the transmitter is to be set up for a maximum flow of 380 liter/sec then
with a vortex flowmeter, it would be possible to reliably measure flow
rates from 380 liter/sec to 10 liter/sec.

One of the features of some flow transmitters is that the user may specify a
"zero cutoff." This is the flow rate below which the measurement value is
not considered to be reliable and thus the transmitter indicates a value of
zero.

> **Zero cutoff** – Flow rate below which the transmitter indi-
> cates a value of zero for the flow measurement.

The ability to specify a zero cutoff value in flow measurement can help
ensure proper accounting when the flow measurement is integrated over

time to determine the amount of material used or produced by the process. Under most circumstances, this action should have no impact on process control since under normal operating conditions the process flow rates should exceed the specified zero cutoff flow rate.

The vortex flowmeter is typically a two-wire device. The term *two-wire* indicates two wires are used to communicate the flow measurement value as a current signal or digital value indicating the flow rate. The same wires are also used to provide the voltage and current needed to power the electronics associated with the transmitter.

> **Two-wire device** – Field device that uses a single pair of wires for power and also to support communication between the device and the control system.

When a two-wire transmitter is used for measurement, there is no need to supply local power as is required for a four-wire transmitter such as the magnetic flow transmitter.

A special circuit design in a two-wire field device enables the device to receive power while using the same pair of wires for communication. To allow a measurement value to be communicated to the control system as a 4–20 mA current signal (which was commonly done before the development of digital transmitters), the transmitter acts as a variable resistance in a circuit powered by the control system's 24 volt DC power supply. For example, if a transmitter is calibrated for a maximum flow of 380 liter/sec then the resistance provided by the transmitter is regulated to provide 20 milliamps when the flow rate is 380 liter/sec. If the minimum flow, say zero or a value below the zero cutoff, is sensed by the transmitter then this is indicated by the current flow being maintained at 4 milliamps. Within the control system, the current signal provided by a field device is measured and then converted to a digital value that may be used for display and control. Alternatively, the measurement may be directly converted into a digital value by the transmitter.

3.3   Flow Based on Differential Pressure

Often the most economical means of measuring gas or liquid flow is based on sensing the pressure drop (differential pressure) across a restriction in the flow stream, so the majority of gas or liquid volumetric flow measurements made in the process industry are made this way. The restriction is created by installing an orifice plate—a plate with a hole smaller than the pipe diameter—through which the stream flows. When the liquid or gas stream flows though the hole, there is a pressure drop. As the flow rate goes up, this pressure drop increases with the square of the flow rate.

Based on this phenomenon, know as Bernoulli's principle, a linear indication of flow rate may be obtained by taking the square root of the differential pressure across the orifice plate.

The manufacturer of the orifice plate sizes the hole based on the expected maximum flow rate, the stream composition, and the operating conditions. The data sheet provided with the orifice plate provides information on the expected differential pressure that will be sensed for standard operating conditions, that is, pressure and temperature. This information is used to set up the differential pressure transmitter to provide a flow rate indication.

Modern digital differential pressure transmitters can be set up to take the square root of the differential pressure and provide an output of flow rate. Alternatively, the processing power of the control system may be used to convert a differential pressure measurement into a linear flow value. However the conversion is made, deviations in the operating pressure and temperature from the normal conditions assumed by the orifice manufacturer may be compensated by also measuring the pressure upstream of the orifice plate and temperature downstream of the orifice plate and using these measurements to calculate a correction factor. An example of flow measurement based on differential pressure and an orifice plate is shown in Figure 3-3.

Figure 3-3. Flow Measurement Based on Differential Pressure and Orifice

When a differential pressure transmitter is purchased the range of pressure measurement is specified based on the expected pressure drop created by the installation of an orifice plate in the pipe. The differential pressure transmitter will be calibrated at the factory according to this pressure range. The calibration of an analog or digital transmitter used for differential pressure measurement may be tested by measuring a known differential pressure. The calibration of a transmitter may also be adjusted to accurately reflect reference input signals.

> **Calibrating a transmitter** – Procedure of adjusting a transmitter to accurately reflect reference input values or a known differential pressure or other process measurement.

A manual procedure involving the adjustment of transmitter bias and gain is required to calibrate an analog transmitter. Hand-held tools or tools built into the control system may be used to check and adjust the calibration of a digital transmitter. An example of a hand-held calibration device is shown in Figure 3-4.

**Figure 3-4. Hand-held Device to Check
and Set Transmitter Calibration**

One of the limitations associated with using differential pressure across an orifice for gas or liquid flow measurement is that the flow measurement turndown is only about five to one. Therefore, a flow rate measurement below 20% of the maximum flow that may be accurately and repeatably measured by the device may be unreliable, which can impact both control and calculations based on this measurement. Since batch processes must operate over a wide range of conditions this limited turndown may not be acceptable and another measurement technique may be more appropriate.

For many continuous processes, this limited turndown in flow measurement is not a problem since the plant normally operates at 80–110% of its design production rate. However, continuous plants must be periodically shut down for maintenance or repairs. During the initial re-start of the process, the flow rates through the process may be substantially lower than during normal production. During these periods of process startup, the flow measurement may not be usable by the automatic control system. When this occurs, the process must be manually controlled by a plant operator adjusting the valves that regulate flow through the process. In making these adjustments, the operator must rely on other measurements such as level and pressure. (A side effect is that during plant startup, it is often necessary to use more plant operators than are required for normal plant operations.) So this is an example of where the measurement device itself limits when the control system may be used to automate the process.

The primary advantages of inferring flow from a differential pressure measurement is that this type of device is relatively inexpensive to install compared to other flow measurement technologies. In many industries, as many as 70–80% of flow measurements are based on differential pressure rather than on magnetic, vortex, or other flowmeter technologies.

3.4   Coriolis Mass Flowmeter

As has been mentioned, in batch processes and in some continuous processes, there is often a need to measure a liquid, gas, or slurry stream flow rate. In some cases, the physical properties of the stream do not allow the flow measurement to be made using a magnetic, vortex, or differential pressure flowmeter. For these difficult applications, a Coriolis mass flowmeter may provide an effective means of measuring mass flow. As shown in Figure 3-5, a drive coil is used with a magnet to produce an oscillation in the Coriolis sensor flow tubes. The coil is energized to keep the tubes vibrating at their natural frequency. Electromagnetic detectors located on each side of the flow tubes produce signals the transmitter uses to determine the mass flow by measuring the phase difference between these signals.

To get best results, special attention must be given to the installation of these devices. Also, the cost of these devices is normally greater than that of alternative means of flow measurement. However, for certain applications the mass flowmeter is the only suitable way to obtain a flow measurement. A Coriolis mass flowmeter may be preferred because of the accuracy of the flow measurement that it provides. Also, the fact that a

Figure 3-5. Coriolis Mass Flowmeter

measurement of density is also provided by the transmitter can be of benefit in some applications where concentration is inferred based on density.

3.5 Pressure Measurement

The measurement of pressure within the process is often critical to maintaining the processing conditions needed for proper plant operation. The pressure within a pipe or vessel may be accurately determined based on a change in capacitance or strain induced in a diaphragm exposed to the process. In most cases, the pressure transmitter is a two-wire device. An example of a pressure transmitter is shown in Figure 3-6.

Figure 3-6. Pressure Transmitter

As illustrated in this example, the expected performance of a pressure transmitter can be expressed in terms of span and rangeability.

> **Span** - the difference between the minimum and maximum output signals of a pressure sensor.

> **Rangeability** - the ratio of the maximum full-scale range to the minimum full-scale range of the flowmeter.

A pressure transmitter may be set up to provide a measurement value in terms of gauge or absolute pressure. Gauge pressure is the pressure referenced to atmospheric pressure. A pressure transmitter set up to measure gauge pressure will provide a measurement value of zero when the process is operating at atmospheric pressure. However, if the pressure transmitter is set up to measure absolute pressure, then for the same conditions the transmitter would indicate approximately 14.7 psi. The exact reading of absolute pressure for this condition depends on the altitude at the plant location and on atmospheric conditions.

When absolute pressure is measured, that is normally indicated in the measurement units. For example, if pressure is measured in pounds per square inch then the units for gauge pressure would be psig but for absolute pressure the units would be psia. Whether a transmitter is set up to provide gauge or absolute pressure measurement depends on the processing conditions that are needed for correct operation. In many cases, a mixture of absolute or gauge pressure may be used on the same process; for example, the pressure of the steam supply may be a gauge pressure measurement while the pressure in a vessel that operates at or below atmospheric pressure may be an absolute pressure measurement.

3.6   Temperature Measurement

A temperature transmitter may be designed to use a sensing element (sensor) that is integral to the transmitter or is mounted remote from the transmitter. The most common sensors in use for temperature measurement are the resistance temperature device (RTD) and the thermocouple. Sensor selection for temperature measurement is based primarily on the temperature operating range that must be supported by the transmitter. Also, the cost of installation and the accuracy requirements become important when either type of sensor may be used at the process operating temperatures. The installation cost of an RTD or thermocouple will typically be higher when the sensor is installed remote from the transmitter.

Transmitters that use an RTD provide the most accurate means of measuring temperature, but their operating range is limited. An RTD consists of a fine wire made of a material such as platinum that provides a known electrical resistance, which changes in a predictable way with temperature. The transmitter determines the resistance of the wire by measuring the voltage developed across the RTD for a given current flow.

To avoid inaccuracies imposed by the resistance of the wiring between the transmitter and the RTD, an additional wire may be used for measurement of the voltage at the sensor. Using this three-wire technique allows the voltage to be measured using a high-impedance device. The best accuracy is achieved by using a four-wire installation, in which two wires are used to provide a current flow to the sensor. The other pair of wires is used to measure the voltage across the sensor. Thus, it is possible to purchase RTD elements that are designed for two-, three-, and four-wire installations. In most cases, a temperature transmitter will support any of these remote-mounted elements.

To protect the sensor from damage, it may be installed inside a thermowell. A thermowell looks like a heavy-duty pipe that protrudes into the process. As the gas or liquid stream passes by the thermowell, the temperature inside the thermowell reflects that of the process stream. To ensure that the measured temperature is the process temperature, the measurement element is mounted to be in physical contact with the thermowell. Even so, the mass of the thermowell slows the response of the measurement to changes in process temperature. An example of a temperature transmitter is shown in Figure 3-7.

Figure 3-7. Temperature Transmitter

The other most common sensor for temperature measurement is the thermocouple. The thermocouple consists of two joined wires made of dissimilar metals that, based on the temperature and the choice of metals, generate a voltage that may be used to calculate the temperature. Thermocouples are made of different combinations of materials and are designated by sensor type. For example, a J-type thermocouple may be used in many common process applications. However, to address the higher temperatures that may be encountered in heaters and boilers, the K type thermocouple may be preferred. Based upon application requirements and manufacturer recommendations, it is possible to select the appropriate thermocouple. In some cases, the setup of the transmitter must be changed to reflect the type of sensor selected. Thermocouples may also be installed in thermowells.

3.7 Level Measurement

There are numerous transmitters, based on different technologies, which may be used to measure level in a tank or vessel. As with other measurement devices, the correct choice depends heavily on the application requirements. The most common way of measuring liquid level is to use a pressure transmitter to measure the pressure head induced by the level.

> **Pressure head** – Pressure induced by a given height (or depth) of a liquid column at its bottom end.

When a tank or vessel is open to the atmosphere, the liquid density and the distance from the liquid surface to the transmitter (presuming correct calibration) determine the pressure that is measured. If the pressure transmitter is located at the bottom of the tank, the level may be sensed under most conditions, that is, from full to empty. However, mounting the pressure transmitter at this point can be a problem if solid material can settle to the bottom of the tank. For this reason, the pressure transmitter may be located at a higher position in the tank and thus level below this point will not be reflected in the level measurement. Another common problem is that the material in tank may plug or coat over the sensor, resulting in a false level indication. Where buildup or plugging is a problem, a "purge" may be installed, which is a small line through which water (in most processes) can be periodically directed at the sensor to keep it clean.

When the level of a pressurized tank or vessel is to be measured using a pressure measurement, the pressure at the bottom of the tank is the combined liquid head plus the gas pressure over the surface. To compensate for the gas pressure, a differential pressure transmitter may be used for level measurement, that is, one side of the pressure transmitter is exposed to the gas and the other to the liquid head. Such installations may be done using a differential pressure transmitter with remote seals.

> **Remote seal** - Flush-mounted diaphragm installed at the process vessel and connected to the differential pressure transmitter using a capillary tube filled with fluid.

Examples of level measurement based on pressure are shown in Figure 3-8.

As we have seen, when pressure or differential pressure is used to infer the level in an open or closed tank, to set up the transmitter to provide an

Figure 3-8. Level Measurement Based on Pressure

accurate indication of tank level, it is necessary to know and account for the density of the liquid in the tank. This is because the pressure associated with the liquid head is not only a function of the distance from the sensor to the liquid surface, but also of the density of the liquid. When someone says, "I had to recalibrate the pressure transmitter to give me the right level," what they're really saying (barring a problem with the sensor or transmitter) is that it was necessary to compensate for a change in the liquid density.

To measure the level of dry material in a tank or the level in a tank of liquid material with changing density, radar level measurement may be appropriate. As its name implies, the level is measured using an electronic beam, similar in principle to that used in a radar installation for detecting planes flying through the sky. An electromagnetic wave is emitted by a radar transmitter (note that in this device, the radar transmitter is part of the sensing element) and is reflected off the surface and back to a receiver. Based on the time difference between transmitted and reflected wave, the device can determine the distance to the surface of the liquid or solid material. An example of a radar level measurement is shown in Figure 3-9.

A high level of technology is incorporated in a radar level device to provide a reliable measurement. For example, setup of the device may allow the user to identify reflections from obstructions in the tank, such as support beams, that should not be taken into account in the level calculation. Even so, the success of these devices for level measurement is very much application-specific. For example, in some cases steam or dust in the air above the liquid or solid surface may prevent getting a good measurement. Manufacturer recommendations and experience with similar applications should be considered in choosing the technology that will be used for level measurement.

Figure 3-9. Radar Level Measurement

3.8   Other Measurement Techniques

This chapter described a brief overview of transmitters that are commonly used for basic measurements in the process industry. Many excellent references have been published that provide a detailed review of the wide variety of measurement and measurement technologies. [1]

References

1.   McMillan, Gregory K. *Essentials of Modern Measurements and Final Elements in the Process Industry: A Guide to Design, Configuration, Installation, and Maintenance*, Research Triangle Park: ISA, 2010 (ISBN: 978-1-936007-23-3).

4

On-line Analyzers

In many cases, the measurement requirement may extend beyond the basic process parameters of pressure, temperature, flow, and level to analytical measurements of properties. Material properties, such as density or viscosity, may be critical to process operation and the successful production of a product, so they must be measured. More complex devices, such as mass flowmeters, may be used to indicate density. Also, the composition of a liquid or gas stream may be of interest. For example, in the control of an ammonia plant, the composition of the process stream may be measured at various points in the process using a gas chromatograph or mass spectrometer. Thus, it is common in most plants to find a variety of analytical measurements in use. In this chapter, we address a few of the analyzers that are found in most industries and illustrate some of the installation considerations and some common limitations.

4.1   Sampling vs. In-situ Analyzers

The installation and maintenance of an analyzer for on-line measurement of a physical property or stream composition can often be quite expensive. In most cases, the sensor must be immersed in the gas or liquid stream to be measured, that is, an in-situ analyzer must be used. Much of the maintenance expense is often associated with the cleaning or replacement of the analyzer sensor. The sensor may be plugged or damaged by exposure to the process stream.

> **In-situ analyzer** – Analyzer that provides a stream property or composition measurement by using a sensor directly immersed in the liquid or gas process stream.

The physical nature of the process stream or the associated operating conditions may not allow the analyzer sensor to be directly exposed to the stream. In these cases, the analyzer must be designed to include a sampling system. The design and installation of the sampling system depends on the type of measurement desired and the process operating conditions. For example, if the stream is at an elevated pressure and temperature, the sampling system may need to provide a sample at some reduced pressure and temperature. Also, if there are particulates contained in the stream, these may have to be removed using filters or other techniques, unless the composition of the stream is to be analyzed.

> **Sampling analyzer** – Analyzer that includes a sampling system to condition a sample of the gas or liquid stream then measured by the sensor.

An example of the sampling system provided with a sampling analyzer is illustrated in Figure 4-1.

The valves and other components such as filters that make up the sampling system of a sampling analyzer often require maintenance on a scheduled basis. Also, depending on the nature of the process stream, the sampling lines and the point at which they enter the process may be prone to plugging. Thus, in some plants, the maintenance of the sampling analyzers is a full-time job done by a special maintenance team. Faults associated with the sampling system can impact the reliability of the analyzer and the use of the analyzer measurements by the control system.

In addition, the physical size of an analyzer and sampling system and need for easy access to do routine maintenance may make it necessary to physically locate the analyzer some distance from the sampling point. Thus, the time required for a sample to travel to the analyzer may introduce a significant delay in the measurement provided by the analyzer. For these reasons, an in-situ analyzer may be a better choice than a sampling analyzer if both types of analyzers are suitable and available for the process measurement.

4.2   Flue Gas O_2

Fuel-fired heaters are commonly used in the process industry to heat a liquid or gas stream. Also, packaged or utility boilers may be used in a plant to meet the steam requirements of the process. When natural gas or fuel oil is burned, a small amount of air over and above that required for the combustion process must be injected with the fuel to provide the mixing needed to ensure all the fuel is consumed. However, if more air is injected

Figure 4-1. Example Sampling System

than is required for complete combustion, this additional air will reduce boiler efficiency.

Air is composed of approximately 21% oxygen, with the remaining components consisting primarily of nitrogen and CO_2. During combustion, any excess air in the combustion chamber is heated to a very high temperature and then discharged with the products of combustion. The energy required to heat this excess air is lost. Therefore, there is a strong economic incentive to maintain the oxygen level measured in the flue gas exiting a boiler or heater as low as possible, but no lower than the level of excess air required for complete combustion. Thus, a measurement of the oxygen content in the flue gas leaving the boiler plays a key role in combustion control.

An in-situ flue gas analyzer may be used to accurately and reliably measure the oxygen content of flue gas. The sensor technology is based on a zirconium oxide cell. One side of the sensor is exposed to atmospheric (ambient) air, and the other side is directly exposed to the gases exiting the combustion process. Based on the difference in oxygen concentration across the cell, a voltage is generated that is used by the analyzer to provide a measurement of the oxygen concentration in the flue gas.

When an in-situ flue gas oxygen analyzer is purchased, the correct probe length must be selected. The length should be selected to allow the sensor end of the probe to be located in the center of the gas stream exiting the boiler or heater. An O_2 analyzer with a relatively short probe is illustrated in Figure 4-2.

Figure 4-2. In-Situ Flue Gas Oxygen Analyzer

Analyzers such as those used to measure flue gas oxygen concentration are designed to allow the device calibration to be checked and the calibration to be adjusted to compensate for variations in the sensor. Connections are provided on the O_2 analyzer that allows calibration gas to be injected into the analyzer sensor to displace the combustion gases. Bottles of calibration gas may be purchased in various concentrations depending on the range the analyzer is used to measure. For example, calibration gas that represents 5% O_2 concentration could be connected to the O_2 analyzer. Periodically or on request, the analyzer cycles through a calibration procedure during which the calibration gas is injected into the analyzer. A check is made by the device to verify the measured O_2 concentration matches the known concentration of the calibration gas.

While the device is being calibrated, the measurement value reflects the O_2 concentration in the calibration gas rather than the actual concentration of O_2 in the flue gas; thus, during this calibration procedure, it is important the control system is aware the measurement does not represent the actual concentration of O_2 in the flue gas. So, manufacturers provide a discrete contact that may be used as an input to the control system to indicate whether calibration is active. This input may be used by the control system to take appropriate action. In this example, when O_2 concentration is used by the control system to adjust the ratio of fuel to air used in combustion, then during O_2 calibration the last measured value should be maintained by the control system and no control action taken based on this measurement.

4.3   Liquid Stream pH and ORP

The probe technology used for a liquid stream pH analyzer is often the most critical part of the measurement. Multiple types are in use. For example, a pH sensitive electrode (usually glass) may be used for pH measurement. Based on the type of probe sensor used, a low millivolt signal may be generated that the analyzer uses to provide a measurement value. In some cases, the transmitter may be calibrated based upon the type of sensor. Thus, the same transmitter may be used to provide different measurements. For example, as illustrated in Figure 4-3, the analyzer may be set up to measure pH or ORP (oxygen reduction potential).

Figure 4-3. pH/ORP Transmitter

Such flexibility in analyzer design and in the selection of probes allows one device to be used to address the needs of a variety of applications. However, considerable expertise may be required to select the probe that best fits the application requirements. Also, the setup and calibration of the device may be more involved than the calibration of a single-purpose device, such as a pressure transmitter. The instrument people in the plant have to be fairly knowledgeable about the measurements taken to keep these devices working correctly. Experience and understanding are required to correctly install and maintain analytical devices, such as a pH/ORP transmitter.

4.4   On-line Estimator

As we saw in Chapter 2, on-line analyzers used to make a sampled or continuous measurement of a physical property or the composition of a liquid or gas stream may not be commercially available, or there may be other conditions that prevent their use. In such a case, the measurements must be obtained by lab analysis of a process grab sample. The time required to collect the grab sample and process it in the lab introduces a significant delay in obtaining the results of the analysis. Often such measurements are needed by the plant operator or control system to maintain key quality-related parameters by making adjustments to plant operating conditions; thus, keeping product properties within specifications. Any delay introduced in processing a sample limits the usefulness of the measurement.

To obtain an immediate indication of the results of process changes that impact a quality-related parameter, it is often possible to use upstream measurements such as flow rate, temperature, and feedstock composition to calculate an estimated value of the quality parameter. For example, a dynamic linear estimator may be created using the calculation tools available in most distributed control systems. Also, some control systems support non-linear estimator blocks based on neural network technology. [1] When a linear or non-linear estimator is available in the on-line control system, a collection of lab analysis is used in conjunction with a history collection of upstream measurement conditions to create the estimator. Once the estimator has been created, the lab analysis is used to verify the accuracy of the estimated value and to make corrections to compensate for unmeasured changes in the process that impact the property being estimated.

References

1.   Blevins, Terrence L., McMillan, Gregory K., Wojsznis, Willy K., and Brown, Michael W., *Advanced Control Unleashed: Plant Performance Management for Optimum Benefit*, pages 261–306, Research Triangle Park: ISA, 2003 (ISBN: 978-1-55617-815-3).

5

Final Control Elements

The flow of gas and liquid streams through a process may be automatically or manually adjusted using one or more field devices called *final control elements*. In this chapter, we provide an overview of some of the most common final control elements used in the process industry.

5.1 Regulating Valves

A regulating valve is the most widely used final control element in the process industry for the regulation of liquid or gas flow. We often think of a valve as a single unit, but in fact it is made up of multiple pieces as illustrated in Figure 5-1. The liquid or gas stream flows through the main part of the valve, known as the valve body. Within the valve body, the flow is regulated by placing a variable restriction in the flow path. The selection of materials and the construction of the valve body depend on the composition of the flow stream and its operating pressure and temperature.

Although some regulating valves are manually operated, a valve actuator, regulated by the control system, is most often responsible for adjusting the restriction in the valve body. A positioner may be used to transmit valve position information to the control system. The size of the valve and its operational and installation requirements determine the type of actuator that is appropriate. For a very large valve, an electric motor or hydraulic cylinder may be used. However, for most valves used in the process industry, a pneumatic driven actuator is the most commonly applied actuator. Approximately 90 percent of all actuators in service today are driven by compressed air. When compared to the cost of electromechanical and electrohydraulic actuators, pneumatic actuators are relatively inexpensive. as well as easy to understand and maintain. [1]

67

Valve actuator – Device that produces linear or rotary motion in a valve using a pneumatic, electric, or hydraulic source of power.

Located within a pneumatic actuator is a large diaphragm mechanically linked to a metal rod or shaft called the *valve stem*. When air under pressure is injected on one side of the diaphragm, a force is developed that in turn produces a force on the valve stem, causing it and the restricting element to move. On the other side of the diaphragm is a spring. The purpose of this spring is to return the valve to a full open or closed position if air pressure to the actuator is lost. Because of the large surface area of the actuator diaphragm, even a slight change in air pressure will cause a significant change in the force exerted on the valve stem.

Valve stem – Shaft used to move the variable restriction and thus change the flow through a valve.

The coupling between the valve body and the actuator, through which the valve stem passes, is known as the valve bonnet. The bonnet provides a leakproof closure for the valve and contains a wearable material called *packing* that is used to maintain a seal between the bonnet and the stem. The type of packing material selected is based on the pressure, temperature, and composition of the flow stream. Graphite is often preferred as a packing material due to its low friction coefficient and its ability to withstand a wide range of material compositions and processing conditions.

Valve bonnet – Component that provides a leakproof closure for the valve and contains a packing used to maintain a seal between the bonnet and the stem.

The bonnet used with smaller valves may screw directly into the valve body, whereas on larger valves, the bonnet often bolts to the valve body. By tightening the bonnet, the packing may be forced into closer contact with the stem, improving the seal, but at the cost of greater force being needed to move the valve stem. An example of a sliding stem valve with actuator and positioner is shown in Figure 5-1.

In some cases, the valve may be equipped with an electro-pneumatic transducer rather than a positioner.

Electro-pneumatic transducers – Device that converts current or voltage input signals to proportional output pressures.

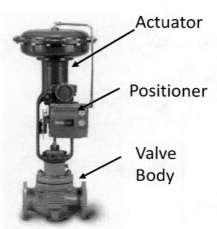

Actuator

Positioner

Valve
Body

Figure 5-1. Components of a Sliding Stem Valve

As mentioned, changes in valve stem position are used to change the degree of restriction within the valve body. There are numerous means that may be used to introduce a variable restriction within the valve body. [2] They may be categorized as disk or plug. With one type of sliding stem valve, the globe valve, the valve stem is coupled to a valve plug that restricts the flow through an orifice in the valve body. Forcing the valve closed causes the plug to make contact with the valve seat, stopping flow.

> **Valve seat** – Replaceable surface used to provide consistent flow shutoff.

A disk or ball valve is used in many applications and introduces a variable restriction through the rotary motion of a disk or ball within the valve body. The ball may have a machined hole or a V-shaped notch. The disk is formed more like a flat plate. The flow restriction introduced by the valve may be adjusted by turning the valve stem and rotating the ball or disk. Thus a disk or ball valve is also commonly referred to as a rotary valve.

> **Rotary valve** - A type of valve in which the rotation of a passage or passages in a transverse plug regulates the flow of liquid or gas through the attached pipes.

One example of a rotary valve is shown in Figure 5-2.

To be able to precisely adjust flow through a regulating valve, it is important that the movement of the valve stem follows the target position set by the control system. A valve positioner is used to regulate the valve actuator to maintain the stem position at the target position provided by the control system. This target position may be communicated to the valve

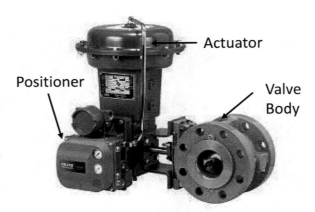

Figure 5-2. Example of a Rotary Valve

positioner digitally or as a 4–20 mA or 3–15 psi pneumatic signal. For example, an input of 4 mA might indicate the target position is closed, whereas a 20 mA input would indicate full open. In between full open and full closed, valve stem position is linear with input value, so in this example, 50% of the current signal range, 12 mA, would indicate a target position of 50% open. It must be kept in mind that 50% open does not necessarily mean 50% of maximum flow rate.

> **Valve positioner** – Device used to regulate the valve actuator to maintain the stem position at a target value.

Depending on whether the control system is pneumatic, analog, or digital, a valve positioner used with the control system may be purely pneumatic or a combination of electronic and pneumatic. A high pressure (e.g., 50 psi) air supply line is provided to the positioner. Many manufacturers of positioners provide electronic or mechanical pressure gauges on the front of the positioner to indicate the supply of air pressure to the actuator. Also, an indication of the target and actual stem position is normally provided. A mechanical link to the stem, or a position sensor coupled to the stem, is used by the positioner to sense stem position. If the actuator is pneumatic, the positioner automatically adjusts the actuator air pressure to maintain the stem position at the target value.

One important decision in choosing and installing a valve is the position the valve will move to on loss of air or hydraulic pressure or loss of connection to the control system. When a positioner is not used, and depending on the actuator design, the valve will remain at its last position or will (more commonly) go to the full open or full closed position, that is, fail open or fail closed. When a positioner is used, it may be set up to take some specific action on loss of communication with the control system.

For example, if power to the control system is lost or the wire from the control system to the positioner is cut, then the positioner may cause the actuator to move the valve to a full open or closed position. The selection of the actuator and positioner setup on loss of power or air supply or loss of communication with the control system is dependent on the process in which the valve is installed. The decision is based on an evaluation of which valve position will minimize or prevent equipment damage while maintaining a safe operating environment. This position is known as the failsafe position.

> **Failsafe position** – Position the valve will revert to on loss of connection to the control system or loss of actuator power supply.

For example, if the steam flow to a heater is adjusted by a valve, then the control system designer must consider which failsafe position would minimize damage to the heater while maintaining a safe operating environment. If the valve failsafe position were selected as fail open, on loss of power supply or connection to the control system, the maximum steam flow would be applied to the heater. However, the material processed by the heater might be ruined by high temperature, or an unsafe condition could be created. Instead, for this example, the failsafe position might be selected as fail closed so the steam flow is cut off on loss of power supply or connection to the control system. This principle also holds in the event of failure of a different control element in the process. In boiler combustion control, the feedwater valve failsafe position might be selected as fail closed in the event of combustion failure to avoid water being introduced into the steam header and risking damage to the steam powered turbines.

As has been mentioned, a valve positioner can be set up to support the desired failsafe position. For example, when the connection to the control system is via a 4-20 mA signal, then the positioner may be configured to interpret 4 mA as either fully open or fully closed. If a valve is set up to fail closed, then the positioner would be set up to close at 4 mA and with increasing signal to open until at 20 mA the valve is full open. On loss of current signal, the valve will go to the closed position. However, if the valve positioner is set up to fail open then the valve will be fully open for a 4-mA signal and fully closed for a 20-mA signal. Therefore, on loss of connection to the control system, the valve will go fully open. Similarly, the valve actuator may be selected to provide fail open or fail close action on loss of air supply. To provide consistency in plant operation, the same failsafe position should be provided for loss of power to the valve actuator as for loss of connection to the control system.

As seen, if the failsafe position is fail closed then an increasing signal from the control system will open the valve. If the failsafe position is fail open then an increasing signal will close the valve. To allow the operator and the control designer to think and work in terms of implied valve position, most control systems provide an "Increase Open/Increase Close" option. When the option for Increase Close is selected, the signal communicated to the positioner is equal to (100% − implied valve position). Thus, a request by the operator or control system to close the valve (i.e., the implied valve position is 0% open) would provide a signal of 4 mA for an Increase Open selection but a signal of 20 mA would be provided for an Increase Close selection. When such an option is provided by the control system, the operator and control system views and accesses the valve in terms of implied valve position. For example, when the operator changes the indicated valve position to 0%, this means the valve is fully closed.

> **Increase Open/Increase Close Option** – Control system feature that supports valve failsafe setup and permits the operator and control system to work in terms of implied valve position.

Therefore, the increase open/increase close option must be correctly configured in the control system setup to allow the operator and control system to work in terms of implied valve position.

An important aspect of valve design and selection is the "valve characteristic." The restrictive element (plug, ball, disc, etc.) used to regulate flow through a valve may be shaped in different ways that influence the flow rate through the valve as a function of the stem position. Therefore, one of the options in purchasing a valve is the valve characteristic, which is defined as the change in flow rate through the valve with a given change in stem position, based on a constant pressure differential across the valve. The most common options manufacturers offer for valve characteristics are quick opening, linear, and equal percentage.

> **Valve characteristic** – Manner in which flow through the valve changes as a function of stem position for a constant pressure differential across the valve.

The capacity of a valve is often expressed in terms of C_V where C_V is the number of gallons of water that would flow in one minute for a one psi pressure drop across the valve. When a valve is to be purchased, the maximum flow requirement may be specified in terms of C_V for the full open position. The manner in which flow varies with stem position is shown in Figure 5-3 for valves with quick opening, linear, and equal percentage characteristics.

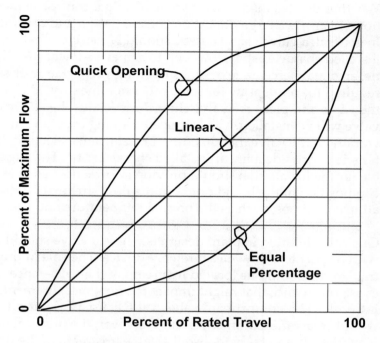

Figure 5-3. Valve Characteristic

In selecting the valve characteristic, it might at first appear that a linear valve characteristic should always be selected if the objective is to obtain the same change in flow for any change in valve position in the valve's operating range. However, it is important to remember valve characteristics are defined in terms of flow for a constant differential pressure. In many applications, the pressure differential across the valve will change with valve position, impacting the installed characteristic, that is, the flow rate as a function of stem position when the valve is installed in the process.

> **Installed characteristic** – Manner in which flow rate changes with valve stem position as installed in the process.

For example, a linear-characteristic valve may be used in combination with a centrifugal pump that establishes the pressure upstream of the valve. The combination of the valve with the centrifugal pump does not result in a linear installed characteristic since the pump discharge pressure changes with flow rate. Thus, the change in flow rate for any specific percent change in the valve stem position depends on the flow rate. However, when a centrifugal pump is used in combination with a valve having an equal percentage characteristic, the installed characteristic is linear.

Whether the installed characteristic of a final control element is linear or non-linear has a large impact on control design and control performance. In general, a valve characteristic should be selected that provides a linear installed characteristic. However, to achieve a linear installed characteristic, an equal percentage or quick opening valve characteristic may be required to compensate for some other nonlinearity. The person who sizes the valve and picks the valve characteristic must be aware of where the valve will be installed. The engineer working on a new installation may work with the valve manufacturer in sizing and selecting the valve characteristic to provide a linear installed characteristic. The engineer responsible for the installation should provide information, such as the composition of the liquid or gas in the flow stream, and the process operating conditions, such as the normal temperature and pressure.

One consideration in plant design is where a valve should be installed with respect to a flow measurement device. The rule of thumb is the flow transmitter should be located upstream of the valve since the valve can cause turbulence that might impact the flow measurement. Turbulence introduced by an upstream valve could cause pressure variations that would create measurement noise. However, the physical layout of piping may make it necessary to install a valve upstream of the flow measurement.

For various reasons, a valve may be installed without a positioner. In this case, the valve will usually be fitted with an electro-pneumatic transducer, an "I-to-P," that takes the 4–20 mA current signal from the control system and, based on the signal value, provides a 3–15 psi signal to the valve actuator. The cost of an I-to-P transducer is much less than the cost of a valve positioner. In such an installation, there is no means to automatically adjust the air pressure to the actuator to ensure the required position is maintained by the valve. Consequently, it is common to have large offsets between the target valve position requested by the control system and the actual valve position. This offset is often caused by the force needed to overcome the friction of the valve packing. The valve manufacturers' guidelines on when to use a positioner versus an I-to-P transducer have changed dramatically over the last 40 years. In the early 1970s, many manufacturers of valves recommended valve positioners be used with only a limited number of applications. Plants constructed in that timeframe may still have valves that do not have a positioner. However, based on field experience, the recommendation today is a positioner should always be used with a valve. There are very few installations that would not benefit from all valves being installed with a positioner. An example of a valve with an electro-pneumatic transducer is shown in Figure 5-4.

Figure 5-4. Valve with Electro-Pneumatic Transducer

When an I-to-P is used with a valve, it is necessary to provide an air supply line. For example, the supply pressure might be 50 psi. In most cases,
the input from the control system to the I-to-P is a 4–20 mA signal, and the
I-to-P might be set up to provide 3–15 psi to the actuator as the control system input varies from 4 to 20 mA. However, the I-to-P has no means of
automatically adjusting actuator pressure to maintain the position
requested by the current signal from the control system since it does not
use a measurement of stem position. If the valve is sticking, then it may
not move as requested in response to a change in the control system input.
Therefore, a positioner should be installed if the valve is to be used in a
control application.

5.2   Damper Drives

The air requirements associated with a boiler are achieved using fans and
large ductwork to transport the air to the boiler. To regulate the flow of air
to the boiler, a large damper may be installed in the ductwork. These
dampers are similar to those found in a home heating system but on a
much larger scale, for example, 15–20 feet in height. A damper drive may
be installed with the damper to allow its position to be regulated by the
control system. An example of a damper drive is shown in Figure 5-5.

A damper drive must exert enough torque to overcome the mass of the
damper and the force exerted by the pressure of moving air on the
damper. Damper drives may be powered by electric or pneumatic supply
lines. The rotary motion provided by the damper drive is transmitted to
the damper using metal linkages. In an application fitted with a damper

Figure 5-5. Damper Drive

drive, the linkage may introduce deadband and hysteresis [3] in the air flow response seen for a change in the control system output to the damper drive.

> **Hysteresis** – The difference in the input signals required to produce the same output during a single cycle of input signal when cycled at a rate below which dynamic effects are important.

> **Deadband** – Range through which an input signal may be varied, upon reversal of direction, without initiating an observable change in output signal.

In most cases, this problem can be corrected by adjusting the linkage to take out any variation introduced by worn linkage joints or in the pins attaching the damper arm to the damper. Good maintenance of the linkage is important when a damper drive is manipulated by the control system.

5.3   Variable Speed Drives

The flow rate of a liquid or gas through a process may be adjusted by changing the speed of the drive motor for a pump or fan instead of introducing a variable restriction into the flow path. A variable speed drive may be installed with a motor to allow the motor speed to be adjusted. Adjusting the speed of a motor is much more energy efficient than installing a constant speed motor and a valve to vary the restriction in the flow line. Also, the rate of response and accuracy to which a motor may be regulated is equal to or better than that achieved using a regulating valve and a constant speed motor. However, the added cost of using a variable speed motor can be significant. An example of a variable speed drive is shown in Figure 5-6.

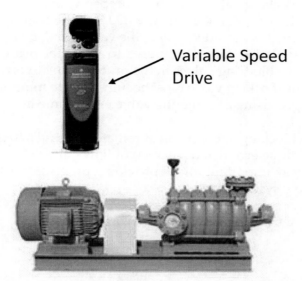

Figure 5-6. Variable Speed Drive and Pump with Electric Motor

When a variable speed drive is used with a pump, it is necessary to include a blocking valve to provide tight flow shutoff when the motor is stopped. The addition of an actuator to allow the blocking valve to be automatically closed by the control system can also increase the cost of an installation. In addition, the range of motor speed adjustment is limited and thus a variable speed drive may not provide the range of speeds necessary to achieve low flow rate targets.

5.4   Blocking Valves

During normal process operation, there may be a need to block off or isolate portions of a process. Thus, blocking valves may be located in the process lines to allow an operator to shut off the flow through the process line to be blocked. For example, to replace a valve, it is necessary to block the line to and from the valve, or in batch applications, automated blocking valves may to used select the source of a feed stock or determine where the product is to be pumped to. Blocking valves also play an important role in start-up. When a process is being started up, blocking valves that were closed when the process was not running must be re-opened at the appropriate time.

A number of variations of blocking valves are available to address different process requirements. In some designs, a blade rather than a plug moves vertically within the valve body to stop flow. However, blocking valves are not designed to be used to regulate a flow stream and will be damaged if they are used to regulate flow.

Although blocking valves may be equipped with on/off, open/closed actuators, these valves often are not tied into the control system and can only be opened or closed locally at the valve using manual handwheels. To allow regulating valves to be used to adjust the process flow, it is important the blocking valves in the process line be correctly set up. For example, the blocking valve must be fully open to minimize any impact on flow achieved using a regulating valve in the same line.

The use of blocking valves in an automated control system is primarily limited to the selection and isolation of flow paths. Automatic blocking valves play an important role in providing process flow shutoff associated with a safety system.

Examples of blocking valves are shown in Figure 5-7.

Figure 5-7. Blocking Valves

References

1.   Skousen, Philip. *Valve Handbook* – Second Edition, Chapter 5.3.1, McGraw-Hill Handbook (ISBN: 0071437738).

2.   McMillan, Gregory K. *Essentials of Modern Measurements and Final Elements in the Process Industry*, Chapter 7-3, Research Triangle Park: ISA (ISBN: 978-1-936007-23-3)

3.   Choudhury, M. A. A. Shoukata, Thornhill, N. F.b, and Shah, S. L. "Modelling Valve Stiction," *Control Engineering Practice*, Vol. 13, pp. 641–658.

6

Field Wiring and Communications

Field devices are most commonly accessed by the control system using cable that is designed to protect the wiring from physical damage and to also provide a shield against stray electromagnetic signals. The types of cable used to connect field devices to the control system may vary because of the standards established by the country in which the plant is located. For plants located in the United States, "twisted pair with shield and drain wire" is the cable traditionally used for wiring field devices for measurement and control. However, some variations in this cable are needed to meet special requirements associated with certain industries. For example, in the oil fields there may be a requirement for armored cable. There are also different ways to wire to field devices; thus, in this chapter, some basic points on wiring installation are addressed.

The need for field wiring largely goes away when wireless transmitters are used for field measurements. In this case, the field devices are accessed by the control system through high-speed wireless communication. This chapter also provides information on the manner in which wireless transmitters are linked into the control system.

6.1 Traditional Device Installation

The traditional way of wiring is to use a cable that contains a "twisted-pair" conductor. The wire used for field wiring can be fairly light-gauge (e.g., 20-gauge wire), since it only needs to carry a small amount of current. A continuous shield, which in most cases is metal foil, is placed around the two wires to minimize electromagnetic interference. A small ground wire known as a drain wire is normally located under the shield to ensure continuity of the shield. The two conductors are twisted to avoid pickup of AC current and other electrical noise, so wiring manufactured in this manner is referred to as twisted pair.

A traditional wiring installation is shown in Figure 6-1. In the case of two-wire field devices, the twisted pair is used to provide the power to operate the device as well as to communicate with the control system using a 4–20 mA signal. For example, the control system analog output cards may communicate the desired valve position over the twisted pair to the valve positioner as a 4–20 mA signal.

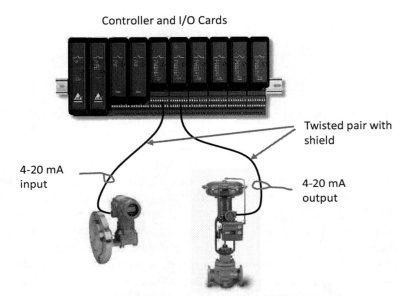

Controller and I/O Cards

Twisted pair with shield

4-20 mA input

4-20 mA output

Figure 6-1. Wiring for Traditional Installation

As previously mentioned, some field devices such as magnetic flowmeters require a considerable amount of power to provide a measurement and are designed as four-wire devices. In this case, the twisted pair going to the device is only used to communicate the measurement value as a 4–20 mA signal. The energy required by the transmitter is provided by a local source of power, typically 110 volts AC. Many on-line analyzers are also designed as four-wire devices, as illustrated in Figure 6-2.

The standard design for analog input cards used in most traditional control systems assumes the field device has no reference to ground, and so the circuit powered by a card may be referenced to the control system ground. Such an assumption is true for two-wire transmitters that are powered by an analog input card. Care must be taken, however, with installing a four-wire device since, in some cases, the 4–20 mA signal provided by the device is not electrically isolated from the electric circuit

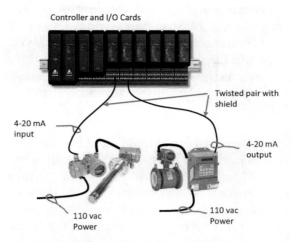

Controller and I/O Cards

Twisted pair with shield

4-20 mA input

4-20 mA output

110 vac Power

110 vac Power

Figure 6-2. Wiring for Four-wire Transmitters

powering the device. The device product literature should specify if the four-wire device output is electrically isolated. In most cases, new field devices are designed to provide signal isolation. However, this may not always be true, especially with older devices.

When the output of a four-wire transmitter is not electrically isolated, connecting the transmitter to an analog input card that is designed to work only with two-wire transmitters can create a ground loop between the field device and the control system. At minimum, a ground loop can impact the accuracy of the measurement indicated by the 4–20 mA signal. In some cases, a ground loop introduced by an input not electrically isolated can disrupt the operation of the controller and affect the operation of other analog input cards installed in the same controller. To avoid these problems, most manufacturers offer isolated analog input cards. However, if such cards are not available, or have not been installed in the control system, it is necessary to install special devices known as *signal isolators* between the field devices and the control system.

6.2   HART Device Installation

Manufacturers introduced digital transmitters to the process industry in the mid-1980s, and sometime later introduced digital valve positioners. These devices were developed by manufacturers to help improve diagnostic information and make it easier to calibrate and setup a field device. Also, an improvement in measurement accuracy could often be achieved. This new generation of transmitters for flow, pressure, temperature, level, and analytic measurement and digital positioners perform their function using microprocessors embedded in the field device. To allow device setup and diagnostics, these digital devices are designed to support bidirectional communication with the control system or with hand-held devices over normal twisted-pair wiring. This communication link may also be used by the control system to read and write field device measurement values and target values. The dominant digital device technology today is known as the HART (Highway Addressable Remote Transducer) Protocol and is an open industry standard supported by the HART User Group. [1]

Communication between a HART field device and a control system over field wiring is fairly low speed, 1200 baud, thus reading or changing a value in HART transmitters and positioners cannot be done very quickly. To allow faster control system access to the device, HART transmitters and final control elements are designed to support a 4–20 mA interface to communicate a measurement or receive a target value. Through this current link, it is possible for the control system to update measurements or provide target values to some devices as often as every 50 milliseconds. This 4–20 mA interface also allows HART devices to be used in control systems that do not support HART communications since nearly all process control systems support a 4–20 mA interface.

In general, the digital communications associated with a HART device ride on top of the 4–20 mA signal between the field device and the control system as illustrated in Figure 6-3. For the control system to utilize HART communications with a transmitter or a final control element, the analog input and output cards of the control system must be designed to use this communication link. Within the control system, the option may be provided to use the digital value communicated by the device or to use the 4–20 mA measurement input. Control system outputs to HART positioners are often limited to 4–20 mA and the HART communication used for diagnostics, setup, and calibration. This capability may be used to digitally access more than one measurement or parameter of the field device. For example, the pressure as well as the body temperature may be accessed through HART communications with a pressure transmitter.

For slower applications, the digital value may be used to provide added accuracy. That is made possible by directly accessing the digital value provided by the device. One example of this is temperature measurement. By using the digital value, it is possible to avoid error introduced by the transmitter in converting the digital value to a 4–20 mA signal and by the analog input card converting the 4–20 mA signal into a digital value. In temperature measurement and control applications, the delay introduced by HART communication may have little impact on control. However, as will be addressed in later chapters, for faster-changing applications such as liquid flow or liquid pressure, the delay introduced by HART-based digital communication can degrade the quality of control. In these faster

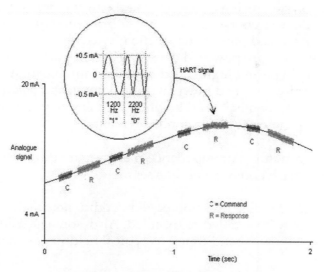

Figure 6-3. HART Communications

applications, the 4–20 mA inputs and outputs should be used to provide the best control.

6.3   Fieldbus Device Installation

As seen, the low communication speed of HART devices may limit the practical use of digital communication to device setup and diagnostics, with the 4–20 mA signal often being used in HART installations to achieve fast updates of transmitter measurements or positioners target. This limitation means a single twisted-pair cable must be installed for each field device. This is "single-drop" installation. In the late 1980s, work was initiated on the ANSI/ISA-50 [2] and the IEC 61158 (International Electrotechnical Commission) standard. [3] Also, work was initiated by other organizations such as PROFIBUS and WorldFIP to develop higher speed communication for use in the process industry. The original objective of the IEC 61158 standard was to define the requirements for digital devices that supported information exchange through high-speed (31.25 Kbaud) digital communications. Also, there was strong interest in reducing installation cost by supporting communications using twisted-pair wiring and to allow multiple devices to communicate with each other and the control system over the same pair of wires, a technique known as *multi-drop installation*.

The dominant digital device technology based on IEC 61158 for use in the process industry is known as Foundation Fieldbus and is an open industry standard supported by the Fieldbus Foundation. [4] In a Foundation Fieldbus installation, it is common for multiple devices to be connected to the control system using a single twisted pair that forms a fieldbus segment. The devices connected over a fieldbus segment can be powered through the twisted pair similar to the way two-wire transmitters and valve positioners are powered over the twisted-pair wires in a traditional installation. To permit high-speed digital communication over the twisted pair, a specially designed power supply, known as a *power conditioner*, is required. Also, for a balanced fieldbus segment, terminators must be located at the far ends of a fieldbus segment.

> **Terminator** - An impedance-matching module used at or near each end of a fieldbus segment.

In some cases, field devices and power conditioners include built-in terminators that may be enabled or disabled. Also, some DCS interface cards incorporate a power conditioner. A typical fieldbus segment is shown in Figure 6-4.

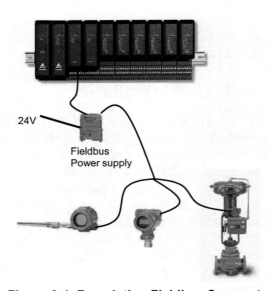

24V

Fieldbus
Power supply

Figure 6-4. Foundation Fieldbus Segment

A control system must use digital communications to access information in field devices that make up a fieldbus segment. Even though the Foundation Fieldbus device communication rate is quite fast compared to HART, these communications cannot be easily synchronized with control implemented in the control system. The resulting delays in utilizing and communicating fieldbus measurement and control actions may impact fast control applications. Thus, to address fast control applications, Foundation Fieldbus devices are designed to support the synchronization of measurement, control and calculations with the communications between devices. This capability to synchronize communications between function blocks contained in field devices is addressed by IEC 61804-2. [5] Also, some manufacturers allow control and calculations to be loaded into the card used for interfacing to Foundation Fieldbus segments. By providing this capability at the card level, it is possible for this control to be synchronized with measurement and positioner targets on the fieldbus segment. Such capability at the interface card level allows control to execute independent of the other control system components and thus can enhance overall system reliability.

6.4   WirelessHART Installation

In an existing installation, the cost to install wiring and electrical power may be a barrier to the addition of new traditional, HART, or fieldbus devices such as PROFIBUS or Foundation Fieldbus devices. It may be possible to address such applications by using battery powered wireless transmitters. The dominant wireless technology used in the process industry is known as WirelessHART. This open industry standard, IEC 62591 Ed. 1.0, is supported by the HART User Group. [1] WirelessHART is also the approved international standard for wireless device installation in the process industry. [6]

WirelessHART devices may be used to address a wide variety of measurement and control applications. When WirelessHART devices are installed, they automatically form a mesh network that may be used to communicate transmitter measurement values, valve position targets, device parameters and diagnostic and setup data to a WirelessHART network gateway. The gateway is connected to the control system to allow measurement values and setup and diagnostic data to be accessed by the control system as illustrated in Figure 6-5. When using WirelessHART devices in control applications, a PID controller designed for wireless control may be required to get the best performance. [7, 8]

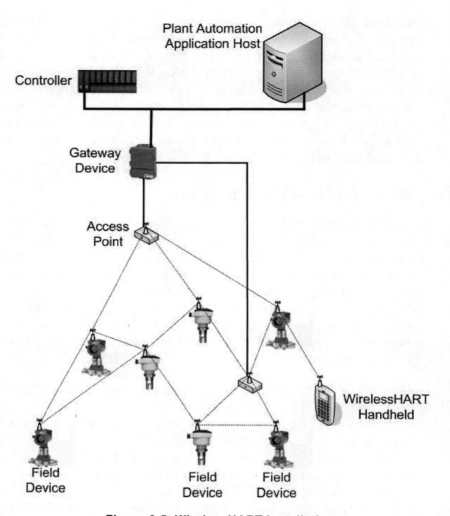

Figure 6-5. WirelessHART Installation

References

1.   HART Communication Foundation, http://www.hartcomm.org.

2.   ANSI/ISA-50.00.01-1975 (R2002) Compatibility of Analog Signals for Electronic Industrial Process Instruments, Research Triangle Park: ISA (ISBN: 1-55617-819-0).

3.   IEC 61158-2 Industrial Communication Networks – Fieldbus Specification – Parts 2 – 6.

4.   Fieldbus Foundation, http://www.fieldbus.org/.

5.   IEC 61804-2 Function Blocks (FB) for Process Control Part 2: Specification of FB Concept and Electronic Device Description Language (EDDL). International Electrotechnical Commission, Geneva, 2007.

6.   IEC 62591 Ed. 1.0 Industrial communication networks - Wireless communication network and communication profiles - WirelessHART™.

7.   Nixon, M., Chen, D., Blevins, T. and Mok, A. "Meeting Control Performance over a Wireless Mesh Network." Paper presented at the 4th Annual IEEE Conference on Automation Science and Engineering (CASE 2008), 23–26 August 2008, Washington, DC.

8.   Chen, D., Nixon, M., Blevins, T., Wojsznis, W., Song, J., Mok, A. "Improving PID Control under Wireless Environments." Paper presented at ISA EXPO 2006, Houston, TX.

7

Control and Field Instrumentation Documentation

To successfully work with (and design) control systems, it is essential to understand the documents that are typically used to illustrate process control and associated field instrumentation. The documentation of process control and associated field instrumentation is normally created by the engineering firm that designs and constructs the plant. The company that commissioned the plant may have an internal documentation standard the engineering firm will be required to follow.

For an older installation, the plant documentation may only exist as a series of paper documents. Today the documentation created for a new or upgraded plant is produced electronically using automated design tools and software. The tools and software selected by the plant or engineering firm for initial plant design or upgrade will influence the documentation format and how documentation is maintained at the plant site. Also, the selection of the control system determines to what extent the system is self-documenting.

> **Self-documenting** – the automatic creation of documents that follow defined conventions for naming and structure.

If the documentation generated by the control system does not follow standards that have been established for process control and instrumentation, then it may be necessary to manually create this documentation.

> **Control System** - A component, or system of components functioning as a unit, which is activated either manually or automatically to establish or maintain process performance within specification limits.

In this chapter, we examine four types of drawings that are commonly used to document process control and associated field instrumentation.

In spite of cosmetic differences, the documentation of process control and field instrumentation for a plant are strongly influenced by and, in some cases, are required to follow standards established for the process industry. For example, companies and engineering firms located in North America may follow standards established by ISA. [1] Accredited by the American National Standards Institute (ANSI), ISA has published more than 135 standards, recommended practices, and technical reports. The standards address control and field instrumentation documentation, as well as other areas such as security, safety, batch control, control valves, fieldbus communication, environmental conditions, measurement, and symbols. Many ISA standards were developed through collaboration with the International Electrotechnical Commission (IEC). The IEC is the world's leading organization that prepares and publishes International Standards for all electrical, electronic, and related technologies—collectively known as "electrotechnology." [2] As previously mentioned in section 2.9, the function block standards, such as IEC 61131 and IEC 61804 (ANSI/ISA-61804), and the batch standards, ANSI/ISA-88 Parts 1-3, have been adopted by many designers of modern control systems for graphics design and documentation of the control system.

7.1 Plot Plan

It is often helpful to look at the plot plan to get an overview of how a plant is physically organized. By examining the plot plan, it is possible to get an idea of where a piece of equipment is located in the plant. A typical plot plan is shown in Figure 7-1.

As will become clear in the following chapters, understanding the physical layout of the plant and the distances between pieces of equipment can often provide insight into the expected transport delay associated with material or product flow between pieces of equipment. For example, how long does it take a liquid, gas, or solid material flow to get from one point in the process to another?

> **Transport Delay** – Time required for a liquid, gas or solid material flow to move from one point to another through the process.

Physically, if the major pieces of process equipment are laid out far apart, then the transport delay can be significant and in some cases, impact control performance. Also, the physical layout of a plant will impact the length of wiring runs and communication distance from the control sys-

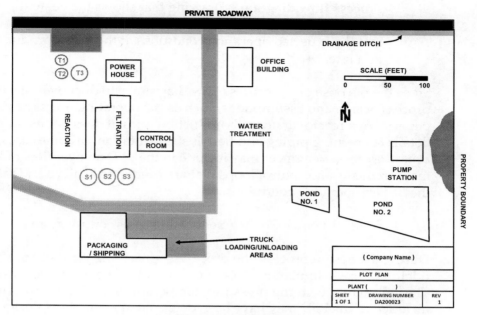

Figure 7-1. Plot Plan

tem to the field devices; thus, it is a good idea to use the plot plan to get a sense of the plant layout and a feel for the location of process equipment and process areas.

7.2   Process Flow Diagram

To meet market demands, a company may commission an engineering firm to build a new plant or to modify an existing plant to manufacture a product that meets certain specifications and that can be manufactured at a specific cost. Given these basic objectives, a process engineer will select the type of chemical or mechanical processing that best meets the planned production, quality, and efficiency targets. For example, if the equipment is to be used to make more than one product then the process engineer may recommend a batch process. For example, a batch reactor may be used to manufacture various grades of a lubrication additive. Once these basic decisions are made, the process engineer selects the equipment that will most cost-effectively meet the company's objectives. Based on the production rate, the process engineer selects the size of the processing equipment and determines the necessary connections between the pieces of equipment. The process engineer then documents the design in a process flow diagram (PFD). The process flow diagram typically identifies the major pieces of equipment, the flowpaths through the process, and the design operating conditions—that is, the flow rates, pressures, and temperatures at normal operating conditions and the target production rate.

Process flow diagram – Drawing that shows the general process flow between major pieces of equipment of a plant and the expected operating conditions at the target production rate.

Since the purpose of the process flow diagram is to document the basic process design and assumptions, such as the operating pressure and temperature of a reactor at normal production rates, it does not include many details concerning piping and field instrumentation. In some cases, however, the process engineer may include in the PFD an overview of key measurements and control loops that are needed to achieve and maintain the design operating conditions.

Control Loop - One segment of a process control system.

During the design process, the process engineer will typically use high-fidelity process simulation tools to verify and refine the process design. The values for operating pressures, temperatures, and flows that are included in the PFD may have been determined using these design tools. An example of a process flow diagram is shown in Figure 7-2. In this example, the design conditions are included in the lower portion of the drawing.

7.3   Piping and Instrumentation Diagram

The instrumentation department of an engineering firm is responsible for the selection of field devices that best matches the process design requirements. This includes the selection of the transmitters that fit the operating conditions, the type and sizing of valves, and other implementation details. An instrumentation engineer selects field devices that are designed to work under the normal operating conditions specified in the process flow diagram. Tag numbers are assigned to the field devices so they may be easily identified when ordering and shipping, as well as installing in the plant.

Tag number – Unique identifier that is assigned to a field device.

The decisions that are made concerning field instrumentation, the assignment of device tags, and piping details are documented using a piping and instrumentation diagram (P&ID). A piping and instrumentation diagram is similar to a process flow diagram in that it includes an illustration of the major equipment. However, the P&ID includes much more detail about the piping associated with the process, to include manually operated blocking valves. It shows the field instrumentation that will be wired

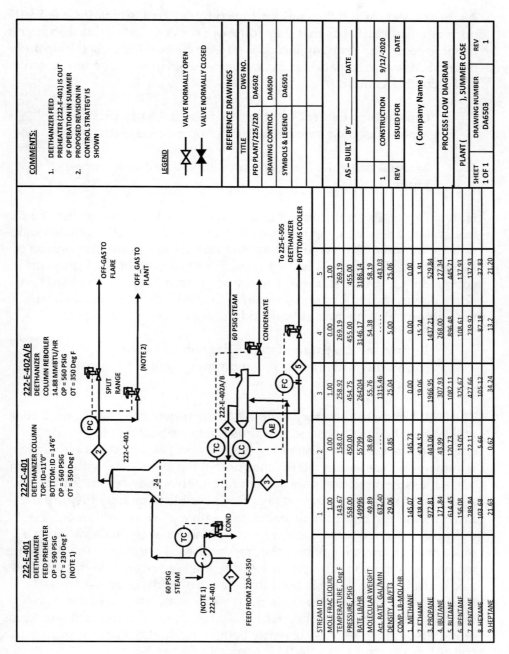

Figure 7-2. Process Flow Diagram

into the control system, as well as local pressure, temperature, or level gauges that may be viewed in the field but are not brought into the control system.

As mentioned earlier, the engineering company that is creating the P&ID normally has standards that they follow in the creation of this document. In some cases, the drawing includes an overview of the closed loop and manual control, calculations, and measurements that will be implemented in the control system.

> **Closed loop control** - Automatic regulation of a process inputs based on a measurement of process output.

> **Manual control** – Plant operator adjustment of a process input.

However, details on the implementation of these functions within the control system are not shown on the P&ID. Even so, the P&ID contains a significant amount of information and in printed form normally consists of many D size drawings (22 x 34 inches; 559 x 864 mm) or the European equivalent C1 (648 x 917mm). The drawings that make up the P&ID are normally organized by process area, with one or more sheets dedicated to the equipment, instrumentation, and piping for one process area.

> **Piping and Instrumentation Diagram** – Drawing that shows the instrumentation and piping details for plant equipment.

The P&ID acts as a directory to all field instrumentation and control that will be installed on a process and thus is a key document to the control engineer. Since the instrument tag (tag number) assigned to field devices is shown on this document, the instrument tag associated with, for example, a measurement device or actuator of interest may be quickly found. Also, based on the instrument tags, it is possible to quickly identify the instrumentation and control associated with a piece of equipment. For example, a plant operator may report to Maintenance that a valve on a piece of equipment is not functioning correctly. By going to the P&ID the maintenance person can quickly identify the tag assigned to the valve and also learn how the valve is used in the control of the process. Thus, the P&ID plays an important role in the design, installation and day to day maintenance of the control system. It is a key piece of information in terms of understanding what is currently being used in the plant for process control. An example of a P&ID is shown in Figure 7-3.

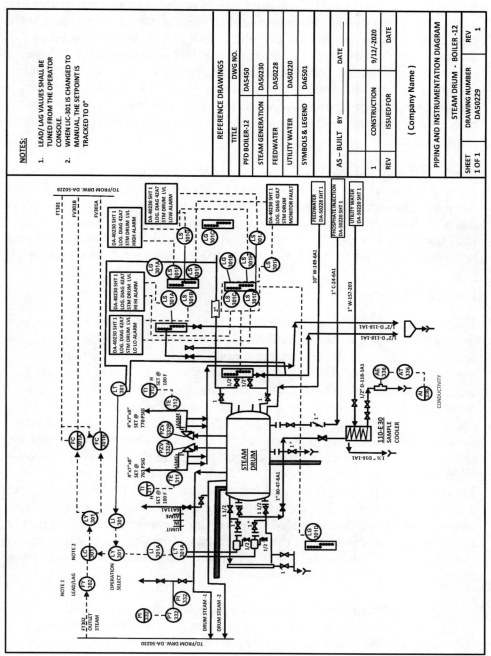

Figure 7-3. Piping and Instrumentation Diagram

When you are doing a survey of an existing plant, obtaining a copy of the plant P&IDs is a good starting point for getting familiar with the process and instrumentation. Unfortunately, the presentation of process control on the P&ID is not standardized and varies with the engineering firm that creates the plant design. In some cases, process control is illustrated at the top of the drawing and its use of field instrumentation is indicated by arrows on the drawing that point from the field instrumentation to the control representation. Another common approach is to show control in the main body of the drawing with lines connected to the field instrumentation. Using either approach complicates the drawing and its maintenance since process control design may change with plant operational requirements.

For this reason, the P&ID may only show the field instrumentation, with other documentation referenced that explains the control and calculations done by the control system. For example, when the process involves working with hazardous chemicals, then a controller functional description (CFD) may be required for process safety management (PSM). Standards have been established by OSHA for controller functional descriptions. [3]

7.4 Loop Diagram

The piping and instrumentation diagram identifies, but does not describe in detail, the field instrumentation that is used by the process control system, as well as field devices such as manual blocking valves that are needed in plant operations. Many of the installation details associated with field instrumentation, such as the field devices, measurement elements, wiring, junction block termination, and other installation details are documented using a loop diagram. ISA has defined the ISA-5.4 standard for Instrument Loop Diagrams. [4] This standard does not mandate the style and content of instrument loop diagrams, but rather it is a consensus concerning their generation. A loop diagram, also commonly known as a loop sheet, is created for each field device that has been given a unique tag number. The loop diagrams for a process area are normally bound into a book and are used to install and support checkout of newly installed field devices. After plant commissioning, the loop Diagrams provide the wiring details that a maintenance person needs to find and troubleshoot wiring to the control system.

> **Loop Diagram** – Drawing that shows field device installation details including wiring and the junction box (if one is used) that connects the field device to the control system.

The loop diagram is a critical piece of documentation associated with the installation of the control system. As has been mentioned, the engineering company that is designing a process normally has standards that they follow in the creation of a loop diagram. These standards may be documented by the creation of a master template that illustrates how field devices and nomenclature are used on the drawing.

The loop diagram typically contains a significant amount of detail. For example, if a junction box is used as an intermediate wiring point, the loop diagram will contain information on the wiring junctions from the field device to the control system. An example of a loop diagram is shown in Figure 7-4.

As is illustrated in this example, junction box connections are shown on the line that shows the division between the field and the rack room. The loop diagram shows the termination numbers used in the junction box and the field device and for wiring to the control system input and output cards. Also, the Display and Schematic portions of the loop diagram provide information on how the field input and output are used in the control system.

Figure 7-4 shows installation details for a two-wire level transmitter that is powered through the control system analog input card. Also, connections are shown between the control system analog output card and an I/P transducer and pneumatic valve actuator. Details such as the 20 psi air supply to the I/P and the 60 psi air supply to the actuator are shown on this drawing. Based on information provided by the loop diagram, we know that the I/P will be calibrated to provide a 3–15 psi signal to the valve actuator. In addition, specific details are provided on the level measurement installation. Since the installation shows sensing lines to the top and bottom of the tank, it becomes clear that the tank is pressurized and that level will be sensed based on the differential pressure.

In this particular installation, the instrumentation engineer has included purge water to keep the sensing line from becoming plugged by material in the tank. Even fine details such as the manual valve to regulate the flow of purge water are included in the loop diagram to guide the installation and maintenance of the measurement device.

In this example, the loop diagram shows the installation of a rotameter. A rotameter consists of a movable float inserted in a vertical tube and may be used to provide an inexpensive mechanical means of measuring volumetric flow rate in the field.

As this example illustrates, the loop diagram provides information that is critical to the installation, checkout, and maintenance of field devices. By examining the loop diagram, it is possible to learn details that may not be

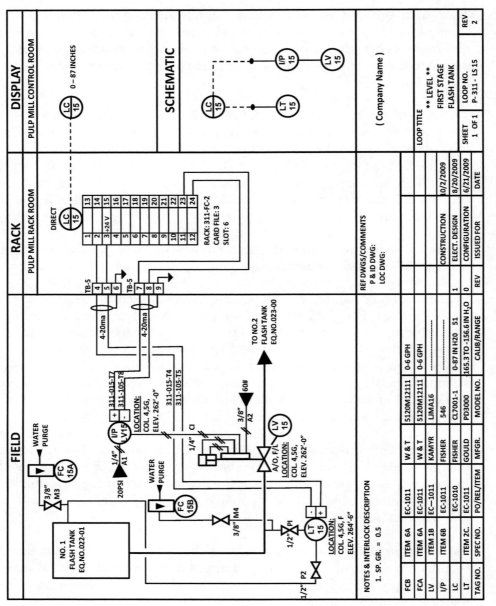

Figure 7-4. Example Loop Diagram—Level Control

obvious when you are touring the plant site. For example, as was previously presented in Chapter 3 on measurement, there are various ways to measure temperature. In the case of a temperature measurement, the loop diagram will provide information on the temperature transmitter, as well as the measurement element that is used. Figure 7-5 shows the loop diagram for a temperature measurement in which a three-wire RTD element is used for the temperature measurement. Details such as the grounding of the shield for the element wire and for the twisted pair going from the transmitter to the control system are noted on the loop diagram.

The process of creating and reviewing loop diagrams is made easier by the fact that many components used to represent measurement and control in similar applications are often repeated. For example, the manner in which the control valve, actuator, and associated I/P transducer were represented in the loop diagram example for a level application is duplicated in other loop diagrams that depict a regulating (control) valve. This is illustrated in the loop diagram shown in Figure 7-6 for a pressure application in which a regulating valve is used in the control of pressure. Also in this example, the pressure measurement is made with a two-wire transmitter. As will be noted by comparing Figure 7-4 and Figure 7-6, the wiring for the pressure transmitter is similar to that used for the level transmitter.

In some cases, the operator uses a process measurement as an indicator or as an input to a calculation that is done in the control system. A loop diagram may be developed for these types of measurement that details the device installation and wiring to the control system. In such cases, the loop diagram will contain no definitions of control. Figure 7-7 shows a flow measurement made by measuring the differential pressure across an orifice plate. As was previously covered in Chapter 3 on measurement, the hole in the orifice plate is sized to give a specific pressure drop at the maximum flow rate that the process is designed to support. As noted in the lower portion of the loop diagram, the orifice plate is sized to provide a differential pressure of 500 inches H_2O at a flow rate of 750 gpm. Also noted in the loop diagram is the control system is expected to take the square root of this differential pressure to obtain an indication of flow rate.

7.5   Tagging Conventions

The tagging conventions that were shown in examples of a P&ID and loop diagrams may be confusing to someone who has not worked with these documents. The naming convention illustrated in the P&ID and loop diagram examples are fairly well standardized in North America. To a certain extent, similar conventions are used in Europe and Asia to document process instrumentation and control. The standard tagging convention used

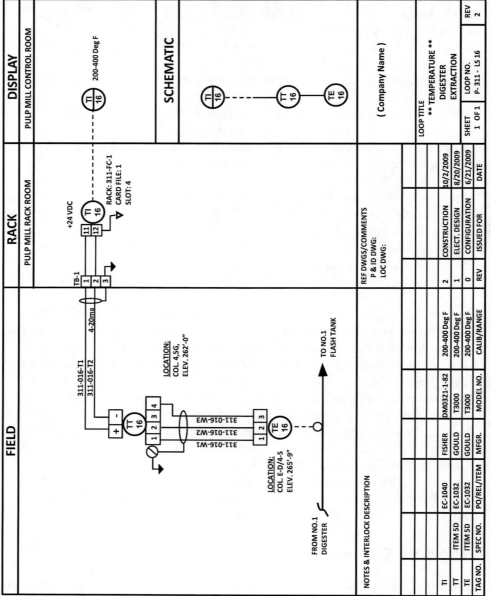

Figure 7-5. Example Loop Diagram – Temperature Control

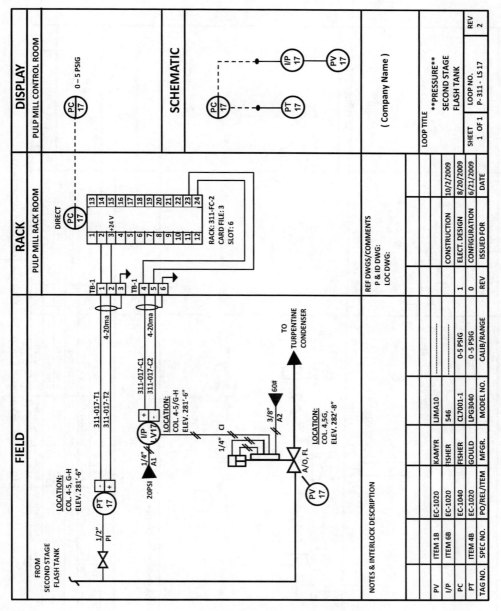

Figure 7-6. Example Loop Diagram – Pressure Control

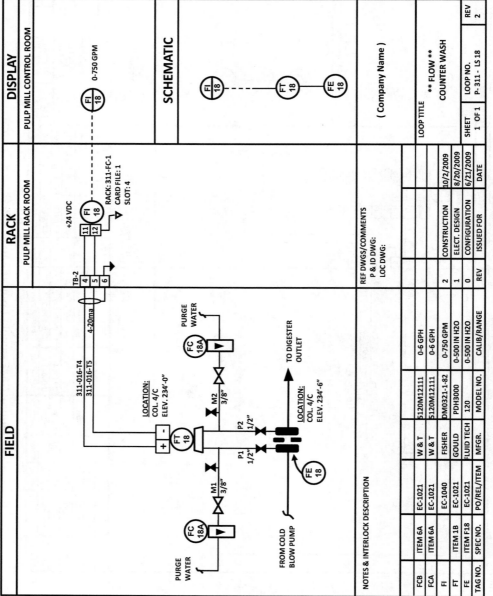

Figure 7-7. Example Loop Diagram – Flow Measurement

in North America has been developed through the efforts of ISA. The format of tags used to identify field devices is defined by ISA-5.1 - Instrumentation Symbols and Identification.

The ISA-5.1 standard defines a standard tag number convention for valves, transmitters, other field devices, and calculation and control functions. In addition, the standard defines expanded tag number conventions that may better meet the documentation requirements of large installations that consist of many process areas or multiple measurements that serve similar functions. The tag number convention defined by ISA-5.1 is summarized in Figure 7-8.

```
                    TYPICAL TAG NUMBER
    TIC 103     - Instrument Identification or Tag Number
    T   103     - Loop Identifier
        103     - Loop Number
    TIC         - Function Identification
    T           - First-letter
    IC          - Succeeding-Letters

                 EXPANDED TAG NUMBER

    10-PAH-5A - Tag Number
    10           - Optional Prefix
               A - Optional Suffix

    Note: Hyphens are optional as separators
```

Figure 7-8. ISA-5.1 Tag Number Convention

The tag number convention established by ISA-5.1 may appear cumbersome, but when examined in detail, it can be easily understood and applied when you are creating new documentation. When first looking at this standard, it is often helpful to consider some examples contained in the standard that illustrate the use of this tagging convention.

The letters that make up the first few characters of a typical tag number (the "leading letters") are used to identify the function performed by the field device or by the control system. Following these leading letters is a number. The number that appears on the tag is known as the loop number. The loop number is used to uniquely identify one or more field devices that are used to perform a specific function. This combination of function letter and loop number allows a field device in a process area to be precisely identified. Knowing the device tag number is required when filling out a work order to or in discussing a field measurement with an operator or instrument technician. The tag number assigned to a field device is normally stamped on a tag that is attached to the device.

All the devices that are used together to perform a specific function are normally assigned the same loop number. For example, the flow transmitter and regulating valve used to measure and regulate the flow of a process stream may be assigned loop number 101.

The loop number normally has only three digits. Consequently, the number of field devices that can be uniquely identified using the standard tag number convention is very limited. For this reason, an expanded tag number convention, defined by the ISA-5.1 standard, is used in the process industry. The expanded tag number convention allows a number to be inserted in front of the function, and that number is usually the process area number. As discussed in Chapter 2, a plant is divided into process areas that are assigned a number. The combination of the area number, the function letters, and the loop number is unique within a plant.

In some cases, multiple field devices may be used to perform a similar function. For example, in some boiler applications the temperature of each tube in the superheater section of a boiler may have individual temperature measurements. Rather than assigning a loop number to each measurement, the expanded tag number convention assigns a common loop number to all these measurements. When this is done, one or more characters may be added after the loop number to uniquely identify each measurement. For the boiler example, loop number 105 might be assigned to all the superheater temperature measurements and the individual measurements may be identified by adding an A after the loop number for the first measurement, a B for the second measurement, and a C for the third measurement.

This option to add letters after the loop number allows unique tag numbers to be created for each measurement, even when a large number of similar measurements is made in a process area. A hyphen may be optionally used in the tag number to separate the area number or characters added after the loop number. However, in general, the use of a hyphen in the tag number is not recommended since in many control systems, the length of a tag number is limited to a maximum number of characters (e.g., 12 or 16 characters).

As defined by ISA-5.1, the identification letters used to specify the function of a field device are organized in a specific manner. The meaning of a letter varies depending on whether it is the first letter or a succeeding letter. A table of the identification letters defined by the standard is shown in Figure 7-9. If the first letter of a tag number is an A, this indicates that the primary function of the device is analysis; if the first letter is an F, then the primary function is flow. When the first letter is H, this indicates that a manual or Hand function is to be performed. By reviewing this table of identification letters, the use of the letters in a tag number can be easily determined.

		First Letters		Succeeding Letters		
		Measured/Initiating Variable	Variable Modifier	Readout/Passive Function	Output/Active Function	Function Modifier
A		Analysis		Alarm		
B		Burner, Combustion		User's Choice	User's Choice	User's Choice
C		User's Choice			Control	Close
D		User's Choice	Difference, Differential			Deviation
E		Voltage		Sensor, Primary Element		
F		Flow, Flow Rate	Ratio			
G		User's Choice		Glass, Gauge, Viewing Device		
H		Hand				High
I		Current		Indicate		
J		Power	Scan			
K		Time, Schedule	Time Rate of Change		Control Station	
L		Level		Light		Low
M		User's Choice				Middle, Intermediate
N		User's Choice		User's Choice	User's Choice	User's Choice
O		User's Choice		Orifice, Restriction		Open
P		Pressure		Point (Test Connection)		
Q		Quantity	Integrate, Totalize	Integrate, Totalize		
R		Radiation		Record		Run
S		Speed, Frequency	Safety		Switch	Stop
T		Temperature			Transmit	
U		Multivariable		Multifunction	Multifunction	
V		Vibration, Mechanical Analysis			Valve, Damper, Louver	
W		Weight, Force		Well, Probe		
X		Unclassified	X-axis	Accessory Devices, Unclassified	Unclassified	Unclassified
Y		Event, State, Presence	Y-axis		Auxiliary Devices,	
Z		Position, Dimension	Z-axis, Safety Instrumented System		Driver, Actuator, Unclassified final control element	

Figure 7-9. ISA-5.1 Identification Letters

In some cases, a letter can only be used as a succeeding letter. For example, the letter D would never be used as a first character, but it may be used as a succeeding letter to indicate "differential." Thus, the combination PD is valid and would indicate the function of the device is "pressure, differential," that is, differential pressure. The combination HIC would be used to indicate hand indicator controller, that is, manual control. Indication and control based on an analytic measurement would be identified as AIC. The letter combination FIC is quite common and used to indicate a "flow indicating control" function. A control valve used in pressure control would be identified using the letters PV, pressure valve. A temperature measurement used only for indication would be identified as TI, "temperature indication." A position transmitter would be identified using the letters ZT.

In the ISA-5.1 standard, several examples are provided that illustrate the correct use of the indication letters within a tag number. Also, the standard covers other letters that may be used in special applications.

The use of tag numbers in a process control system is quite straightforward and must be understood to work with or create documentation for a process control system.

7.6   Line and Function Symbols

Different types of lines are used in process flow diagrams, piping and instrumentation diagrams, and loop diagrams to indicate the type of connection between field devices and the control system. The ISA-5.1 standard defines the instrument line symbols that are commonly used in control system documentation. As illustrated in Figure 7-10, a solid line is used to represent a physical connection to the process. Two slashes shown as points along a line are used to indicate a pneumatic signal. One of the most common ways to indicate an electric signal is a dashed line as defined in ISA-5.1. Communication links between devices and functions of a distributed control system are indicated by small bubbles along the line as illustrated in Figure 7-10.

- Instrument supply
  or connection to process

- Pneumatic Signal

- Electric Variable or Binary

- Communication Link

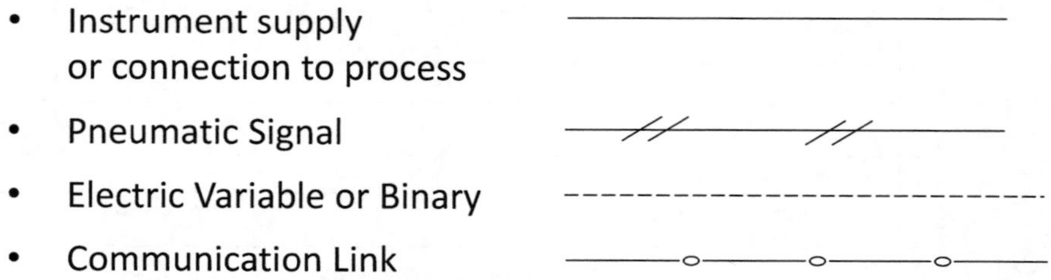

Figure 7-10. Excerpt from ISA-5.1 Instrument Line Symbols

The previous examples of the process flow diagram, piping and instrumentation diagram, and loop diagram contained one or more circle symbols. In these drawings, a circle is used to indicate a discrete instrumentation or control function. A horizontal line drawn through the middle of the circle indicates the function may be accessed by the plant operator. There are many functions, such as those performed by an I/P transducer or valve positioner, that are typically not directly accessible by the operator. Also, some field devices for measurement and actuation may only be accessed by control or calculation functions in the control system and thus would not be shown in the documentation as being directly accessed by the operator. However, the associated control or calculation function that is accessed by the operator would include a horizontal line. As illustrated in Figure 7-11, one of the conventions advocated in ISA-5.1 is to include a square around the circle if the associated function is accessed by an operator through a video display of a distributed control system (DCS). However, in practice, this convention is often not followed.

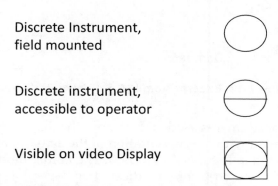

Discrete Instrument, field mounted

Discrete instrument, accessible to operator

Visible on video Display

Figure 7-11. Excerpt from ISA-5.1 General Instrumentation or Symbol Function

It is common practice to illustrate the valve body, as well as the valve actuator and positioner function in control system documentation. The ISA-5.1 standard addresses the representation of a valve body. Most types of valves are addressed by this standard. However, the engineering firm that is designing a process plant may have adopted some variation of what is shown in ISA-5.1. In such cases, it is common practice for the engineering firm to provide a drawing that explains the symbol functions included in their documentation. Also, in some cases, a general valve representation is used rather than different representations for a rotary valve or sliding stem valve. [5] Generally, a damper will be shown rather than the general valve symbol to indicate the regulation of air or gas flow to a boiler or a similar process such as a kiln or heater. An excerpt from ISA-5.1 Valve Body and Damper Symbols are illustrated in Figure 7-12.

Since the type of actuator used with a valve body may impact the operation and failure mode of the valve, the type of actuator is normally indicated in control system documentation. The representation of common types of actuators as defined by ISA-5.1 is shown in Figure 7-13. A complete representation of the valve is provided by combining the valve representation with the actuator representation.

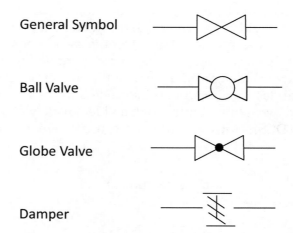

General Symbol

Ball Valve

Globe Valve

Damper

Figure 7-12. Excerpt from ISA-5.1 Valve Body and Damper Symbols

When a positioner is used with a valve, the diaphragm representation may be combined with a representation of the actuator and the valve body, as is illustrated in Figures 7-16 to 7-19. The failure mode of the valve (i.e., fail open or fail closed) is often indicated by the valve symbol in control system documentation.

A special actuator symbol is defined by the standard for motorized actuators. Motorized actuators are used in some industry segments because upon loss of power, the last valve position is maintained. Also, better resolution may be achieved using a motorized actuator for a specific application, such as the basis weight valve that is used to regulate the thick stock flow to a paper machine in paper manufacturing. [6] Solenoid actuators that are used to automate valves used in on-off service are shown using a special symbol.

Installation details are provided in the loop diagram documentation. Some of details may also be provided in the P&ID. For example, the P&ID and loop diagram may show the orifice that must be installed to measure flow using a differential pressure transmitter. Hand-operated valves that are used to block flow during startup or maintenance are shown since they impact process operation if not properly set up. Also, the installation of inline instrumentation, such as a magnetic flowmeter, is commonly shown in a unique manner in the control system documentation. In addition, measurement elements such as an RTD or a thermocouple are shown since they may be physically installed some distance from the field transmitter or the control system. To provide a consistent means of documenting the physical installation, the ISA-5.1 standard includes symbols for many of these common installation details and field devices. A sample of some of these symbols is shown in Figure 7-14.

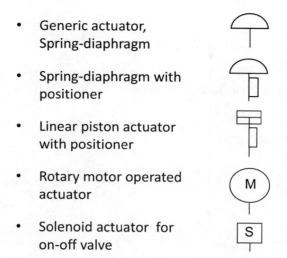

- Generic actuator, Spring-diaphragm

- Spring-diaphragm with positioner

- Linear piston actuator with positioner

- Rotary motor operated actuator

- Solenoid actuator for on-off valve

Figure 7-13. Excerpt from ISA-5.1 Actuator Symbols

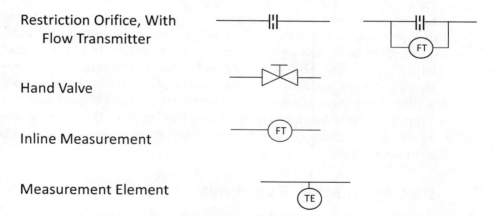

Restriction Orifice, With Flow Transmitter

Hand Valve

Inline Measurement

Measurement Element

Figure 7-14. Excerpt from ISA-5.1 Symbols for Other Devices

7.7   Equipment Representation

A representation of major pieces of process equipment is normally included in control system documentation. This allows the field instrumentation installation to be shown in relationship to the process equipment. Example process equipment representations are illustrated in Figure 7-15.

A general vessel representation may be appropriate for vessels, agitators, heat exchangers, and pumps that do not play an important role in the control system. For example, an agitator on a tank may not directly impact the control associated with the tank level. A special representation is provided

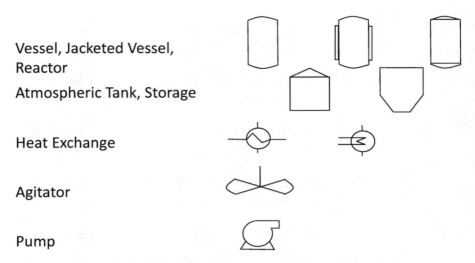

Vessel, Jacketed Vessel, Reactor

Atmospheric Tank, Storage

Heat Exchange

Agitator

Pump

Figure 7-15. Examples of Process Equipment Symbols

for a reactor. A jacketed vessel symbol may be used when a vessel is heated or cooled by circulating liquid through an outside shell. Such a design is commonly used in the batch industry, and permits the vessel's contents to be heated or cooled without coming in contact with the circulating liquid. A symbol is also provided for flat-bottomed and cone-shaped storage tanks that are open to the atmosphere. The examples include a representation of a heat exchanger, which is used to heat or cool a liquid stream. A symbol is defined for an agitator that may be used to ensure good mixing of liquids in a vessel. Also, a pump symbol is shown in these examples.

7.8 Documentation Examples

The four examples included in this section are designed to illustrate many of the recommendations of the ISA-5.1 standard. In the first example, questions and answers are used to highlight specific points that have been covered in this chapter. A review of these questions will hopefully serve to reinforce your understanding of concepts and terminology that will be used in later chapters. By studying the control system and field instrumentation documentation, a lot can be learned about the process and the way the control system is designed to work. Such background information can be extremely helpful when you are working with or troubleshooting the control system for a new process.

7.8.1 Example – Basic Neutralizer Control System

The neutralizer process may be used to adjust the pH of a feed stream. The field devices and process controls that would typically be provided to this process are shown in this piping and instrumentation diagram illustrated in Figure 7-16.

After reviewing this diagram, consider the following questions and answers.

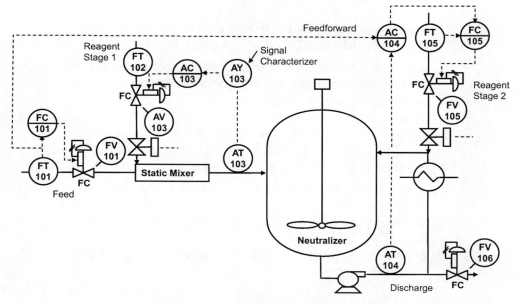

Figure 7-16. Example – Basic Neutralizer Control System

Question: What tag number convention is shown in this example?

Answer: The expanded tag number convention, defined by the ISA-5.1 standard, is not used in this example since the tag only contains function letters and a loop number.

Question: Flow measurements are made using FT101, FT102, and FT105. What type of measurement technique is used by these field devices?

Answer: The field device is shown as installed inline and the liquid is conductive. Thus, a magnetic flowmeter is installed.

Question: Do the valves have digital positioners?

Answer: Yes, an electric signal is shown going directly to a box located on each valve. The pneumatic line from the actuator is also tied to this box. Therefore, we can assume that these valves are installed with a valve positioner.

Question: What type of valve is being used to regulate flow?

Answer: The general valve body symbol is used and thus the type of valve is not indicated in this drawing.

Question: Is the plant operator able to access the control functions indicated by AC103, AC104, FC101, and FC105?

Answer: Yes, the circle indicating these control functions contains a horizontal line that indicates the operator may access this function.

Question: Can the operator access flow measurements FT101, FT102, and FT105, as well as analytic (pH) measurements AT103 and AT104?

Answer: No, the circle indicating these measurement functions does not include a horizontal line. In this case, these measurements are not directly accessed by the operator.

Question: Are any on-off (blocking) valves used in this process?

Answer: Yes, two blocking valves are shown with electric on-off actuators.

Question: What function is provided by AC103?

Answer: The function letters of the tag number indicate that the function is analytic control.

Question: What measurement is used by AC103 in performing its control function?

Answer: Since an electric signal line is shown from AT103 to AC103, then control must be based on the measurement AT103 (pH).

Question: What is the purpose of the function shown between AT103 and AC103, that is, AY103?

Answer: On the diagram, this function is noted as Signal Characterization. When a complex calculation is done, it is often represented by a single function with AY noted inside the circle and a note provided to explain the purpose of this function.

Question: Does the vessel have an agitator?

Answer: Yes, the figure shows an agitator impeller in the vessel, powered by a motor on top of the vessel.

Question: Is the pump fixed speed or variable speed?

Answer: Since control of the process is done using a regulating valve, then it may be assumed that the pump is fixed speed.

Question: How could the liquid in the vessel be heated?

Answer: A heat exchanger is shown in the recirculation line and could be used to heat (or cool) the liquid in the vessel.

Question: How is the pH of the incoming stream adjusted?

Answer: The output of AC103 and AC104 go to a valve and flow loop, respectively, that regulates the flow of reagent to the inlet stream and recirculation line. Thus, it may be assumed that reagent is used to adjust pH.

7.8.2   Example – Basic Column Control

The basic column control system shown in Figure 7-17 is somewhat more complicated than the basic neutralizer control system. However, after studying this example, it will become clear that many of the components that make up the column control system are similar to those used in the basic neutralizer control system. As we have seen, what is learned by working with one process may be often applied to another process. Once you become familiar with a few processes, it becomes easier to work with other processes.

7.8.3   Example – Batch Reactor Control System

Batch processing plays an important role in some industries. In a batch process, a vessel is charged with feed material that is processed in the vessel through mechanical or chemical means (or both). At the end of the process, the product—which may be a finished product or an intermediate product for use in another process—is removed from the vessel. The batch reactor example illustrated in Figure 7-18 is known as "continuous feed batch," in which feed material is continuously added to the vessel throughout part or all of the batch processing. The reaction takes place, and finally the product is pumped from the vessel. In comparing the basic components to those seen in the previous two examples, the only new element introduced by this example is an eductor (a device that produces vacuum by means of the Venturi effect) that is used to remove gases created by the batch reaction.

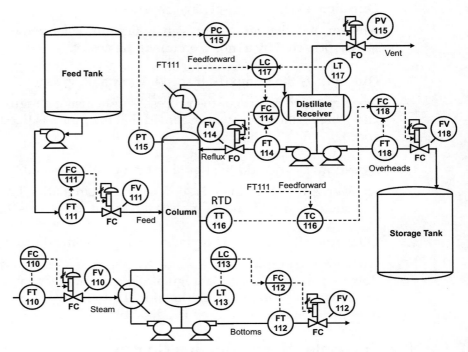

Figure 7-17. Basic Column Control System

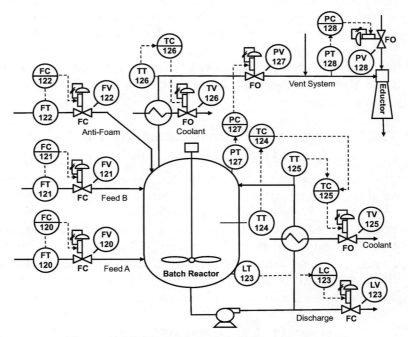

Figure 7-18. Example – Batch Reactor Control

7.8.4   Example – Continuous Feed and Recycle Tank

In studying the example of a continuous feed and recycle tank shown in Figure 7-19, you will note that the only new function that was not contained in a previous example is FFC133. The function letters of the tag number indicate that this new function is ratio control. The line connections indicate that the ratio control is implemented using two feed streams to Reactor 1.

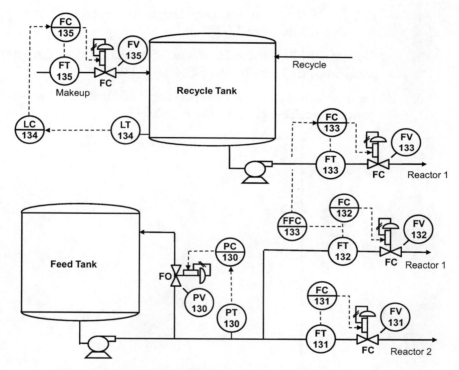

Figure 7-19. Example – Continuous Feed and Recycle Tank

Knowing the tag number and symbols on a P&ID is the first step in understanding which measurement and control functions have been installed on the process. As the examples shown in Figure 7-16 to 7-19 illustrate, a basic understanding of the control and process is possible if you are familiar with the tag number and symbols on the P&ID. Later chapters on control implementation will address how the control functions shown on the P&ID and loops sheet are implemented.

References

1.   The International Society of Automation (ISA). http://www.isa.org.

2.   International Electrotechnical Commission (IEC). http://www.iec.ch.

3.   OSHA Standard 29 CFR 1910.119, *Process safety management of highly hazardous chemicals*. http://www.osha.gov.

4.   ISA-5.4-1991, *Instrument Loop Diagrams*, Research Triangle Park, NC: ISA (ISBN 1-55617-227-3).

5.   Skousen, Philip. *Valve Handbook* – Second Edition, Chapter 1.5.1, McGraw-Hill Handbook (ISBN: 0071437738).

6.   Lehtinen, Marjaana. More speed and accuracy for basis weight control, pp. 22–24, *Automation 1*, 2004.

8

Operator Graphics

In a modern digital control system, the operator's primary interface to the process is made up of one or more CRT/LCD displays (monitors), keyboards, mice, and sound cards. This interface replaces the functionality that was once provided by a control room panel board filled with analog controllers, meters and digital indicators. In addition, functions are now incorporated that were previously provided by the panel motor start/stop buttons and status indicators, chart recorders, annunciator panels and external data sources.

As described in detail in Chapter 2, because of processor and memory limitations and lack of software in early distributed control systems, the operator interface displays consisted of monitors on which preformatted faceplates represented measurement, control, and alarming. In modern digital distributed control systems, the faceplates have been replaced by graphical operator interfaces.

A graphical operator interface consists of a series of displays shown on the operator screens that illustrate the pieces of equipment that make up the process including vessels, piping and measurement values, and station indications from field devices and alarm conditions. The primary advantage of such an approach is that more information may be presented to an operator in an efficient manner. However, the effectiveness of a graphical operator interface depends greatly on the manner in which the graphics displays are structured during system configuration.

Modern digital control systems generally include a variety of standard tools that can be used to create a graphical operator interface. However, since the process equipment and associated field devices are unique to each plant, the development of the graphical operator interface is custom in many respects. The examples and documentation provided by the control system manufacturer often become simply a guide for the customer or system integrator responsible for system configuration. It is common for a plant to develop their own standard for the use of colors and the presentation of information within the graphical interface display.

Process control environments tend to push the capabilities of graphics systems, both hardware and software. All components of the operator interface must be designed to be highly reliable since the operator must stay in control of the process in all circumstances. Interfaces must be designed to provide repeatable behaviors and user interaction and thus help minimize the possibility of human error. For example, the steps required to access process alarms and to acknowledge alarm conditions should be consistent for all process measurements. Fast call-up of process and historical information is required to minimize any delays in providing the operator with information that may be needed to address an abnormal operating condition. For example, a screen will typically be presented and populated with live data within 1 second. The operator environment is designed to be resistant to unauthorized users damaging the application or the data present on the workstation, or gaining unintended access to applications. The degree of access restriction is often configurable, so that trusted, higher skill level users may have greater access to applications and data at the workstation.

Soon after distributed control systems manufacturers introduced support for graphical displays, an American National Standard was developed for Graphic Symbols for Process Displays, ISA-5.5-1985. [1] The stated purpose of this standard was to establish a system of graphic symbols for displays that are used by plant operators for process measurement and control. This was done to promote comprehension by operators of the information that is conveyed through the displays, and to provide uniformity of practice throughout the process industry. The benefit of such standardization is a reduction in operator errors, faster operator training, and improved operator efficiency in working with the control system. Many companies follow this standard, which is suitable for use in numerous industries. However, the standard's two-dimensional line drawings are a mismatch with current commercial workstation technology in that faster computers with graphics card accelerators and vector graphics are supported by most DCS vendors.

Modern control systems include a graphic display editor and a comprehensive library of pre-built display objects with professionally drawn 3D representations of process equipment including pumps, valves, meters, piping, tanks, and other objects. Also, some systems may offer the capability to import graphics from a number of sources, including

AutoCAD®, and Windows® metafiles such as Visio® vector drawings and Windows bitmap (.bmp) and Joint Photographic Experts Group (.jpg) image formats. These graphical components may be added to a display to create a realistic representation of the static elements of a process. Additional elements may be added to the display that show dynamic information. An example of an operator display created using a library of 3D graphics is illustrated in Figure 8-1.

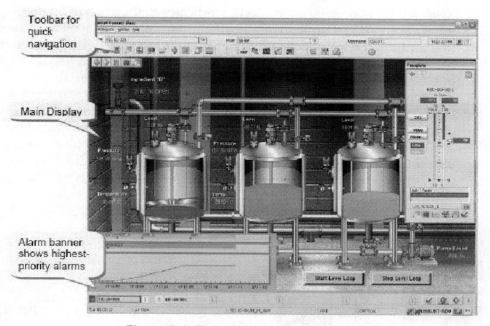

Figure 8-1. Example Operator Display

The representation of the process equipment shown on the P&ID is often taken as a starting point for graphical display design and development, but some judgment must be applied. The reason is these drawings contain many details associated with manual valves and sections of piping that are not reflected in the control system. To create an effective operator display, only information that is important to the operator should be extracted from the P&ID. This usually means that the display will include the measurements, the valves and other final elements, and the control, and then enough of the process equipment and piping so the operator can follow the process flow.

An effective process display allows the operator to quickly find information. For example, if a measurement of interest is associated with a tank, then the operator may quickly recognize the tank on a display and thus readily find the measurement. This means finding information on a

display is typically much faster than presenting a list of items of
information that requires the operator to read each description to find the
information of interest. Additional history integration gives operators a
better perspective on where they have been. Reduced dependency on
numbers through the use of bar graphs, size, etc. allows better operator
scanning of a display.

8.1   Display of Alarm Conditions

Research [2] in user-interface design indicates that alarms in an operator
interface should be implemented using unique shape, as well as color so
that an alarm condition can be recognized more quickly. This approach
has been adopted in the style suggested by alarm standard EEMUA 191
published in 1999 by the Engineered Equipment and Materials Users
Association. [3] Specifically, alarms should be in a unique alarm color.
Pipes, pumps, valves, and so on should not be in alarm colors or in any
other bright color. The implementation of the styles recommended by
EEMUA 191 has greatly influenced the manner in which color is being
used in operator interfaces as illustrated in Figure 8-2 by the difference in
contrast even where shown in black and white, in which these styles were
applied to traditional display design.

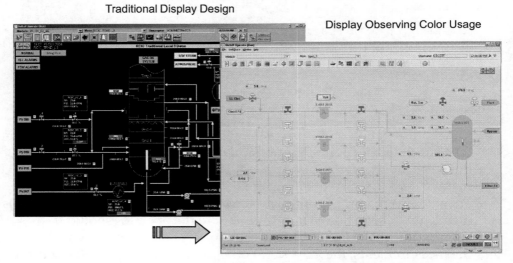

Figure 8-2. Impact of EEMUA 191 on the Use of Color in Displays

Since the publication of EEMUA 191, other standards have been written
that support this work. In 2003 the User Association of Process Control
Technology in Chemical and Pharmaceutical Industries (NAMUR) issued
recommendation NA 102, Alarm Management. [4] Also, the ANSI/ISA-

18.2 - *Management of Alarm Systems for the Process Industries* standard was released in 2009. [5]

It is well known that search times are shortest when the target is significantly different and everything else is similar. Thus, the screen should not be so colorful and busy that the alarm is lost in the confusion.

8.2   Dynamic Elements

Predefined objects for information access are standard parts of any graphical tool set. Basic components, such as knobs, dials, slide bars, and buttons may be dynamically linked to information within the control system. In addition, a library of standard dynamic elements (dynamos) is typically provided that may be used to access measurements, calculation, and analog and digital control components within the control system. Often these pre-engineered dynamos may be customized to fit the specific needs and standards of a plant.

8.2.1   Dynamos

The standard dynamos typically include the key parameters associated with measurement and control functions. For example, a dynamo used to display a control loop may only display the controlled parameter, the setpoint and the controller output. Engineering units provide context to a value displayed by the dynamo. The status of process alarms associated with the control loop is reflected in the dynamo through color change; for example, the background color of the control parameter value. Also, to eliminate clutter in the display, the fact that a loop is not in its designed/ normal mode may be indicated through a color change. For example, if the engineering units are shown in the display as a string, then the background color of the string may be changed to indicate whether the associated control loop is operating in the designed/normal mode.

The ISA-5.5-1985 standard provides general guidelines for color in a display. The predefined dynamos provided with a control system should conform to this standard. However, these guidelines may not be consistent with the manner in which color has been used in a particular plant. In such cases, the colors used in dynamo templates may be modified to conform to the plant standard. An example of the information shown in dynamos that are provided to support the creation of a display is illustrated in Figure 8-3.

When the operator accesses a dynamo, a popup showing a faceplate representation of the referenced measurement or control parameter may be automatically presented that provides further information on the parameter shown. For example, when a dynamo associated with a control loop is

Figure 8-3. Dynamo Examples for Display Building

accessed, the faceplate presented to the operator may be used to modify the control loop mode of operation (manual or automatic) or change a setpoint or output. In addition, the capability may be provided to access further information such as loop tuning by selecting a detail display reference. Parameters in a detail display that may be changed are determined by the allowed user access. In some cases, operator access to information from a graphical display is specified on a parameter by parameter basis. An example of a faceplate display with the associated detail display generated by accessing a control loop dynamo is shown in Figure 8-4.

Just about any parameter in a display can be converted to a dynamic element and animated.

8.3   Displays

The increased use of a mouse in the operator interface has changed the way the operator accesses information within a graphical display. Displays are typically designed to support both horizontal (at the bottom of the display) and vertical (to the right of the display) toolbars. The control

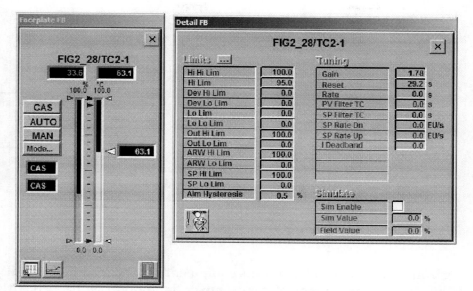

Figure 8-4. Faceplate and Detail Display

system usually provides default toolbars to support the following functions:

- Time and date display
- Alarm list with direct access to the display required to acknowledge the alarm
- Navigation to alarm summary, main menu, system status, last display, page back and forward and print
- Alarm acknowledge, and silencing

Using the display's toolbar to page forward and page back, it is possible for an operator to access displays that contain information upstream or downstream of the displayed process. In addition, dynamos may be added for accessing another display. Through the use of these tools, it is possible to create a display hierarchy that allows access from an overview display to the key display in each process area. Within the process area, it is possible to pan through the displays using the page forward/backward toolbar. Also, using a dynamo for display access, it is possible to provide the operator with navigation between displays that allows fast access to displays that contain startup/shutdown information, displays containing trends of key parameters and in-context documentation, as illustrated in Figure 8-5.

The graphics editor supported by the control system allows the incorporation of animation into a graphical display. Using this capability, the static

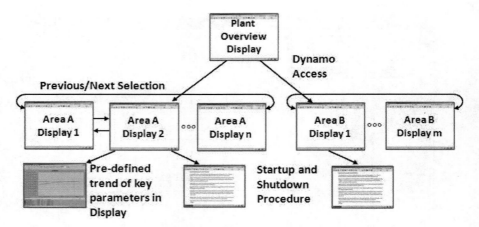

Figure 8-5. Example of Typical Display Hierarchy

graphical component for process equipment may be modified to indicate the status of the equipment, for example, "motor on" or "tripped." Also, animation may be used to represent dynamic data associated with the equipment, such as showing the level of a tank or showing the status of an agitator by rotating the agitator to indicate that the agitator is on. In newer products, dynamos may be provided that are designed to show information in the display based on the state of the process operation.

8.4   Process Performance Monitoring

The graphical display capability of a modern distributed control system may be used to create special screens to show the status of critical equipment. Some examples of the types of information that may be presented in this manner are:

- First Out indication on a process shutdown
- Vibration monitoring
- Boiler soot blower
- Safety system status

The associated display screens are structured to summarize the information. In cases where moving equipment is involved (e.g., a soot blower), animation may be effective in allowing an operator to quickly access which soot blower is active.

The calculation capability of a distributed control system may be used to implement on-line calculation of operation cost, efficiency, and other performance indicators. This type of information may be easily incorporated into the operator's graphical display so that the operator can use this information to improve the process operation. An example of such calculations for a lime kiln operation is shown in Figure 8-6.

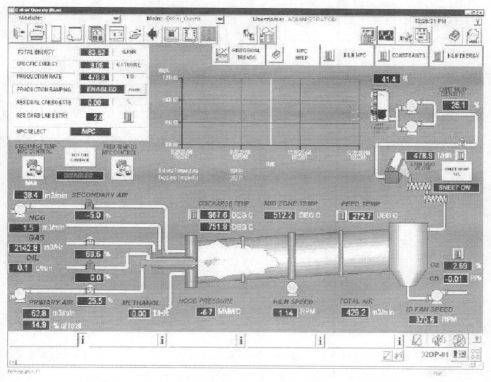

Figure 8-6. Display for Monitoring Performance Indicators

8.5   Process Graphic Data Interfaces

A variety of data sources may be integrated into the process display graphics. The data may be sourced internal to the DCS from the runtime, historians (continuous, batch, event), and alarm services. The data may also be sourced externally from OLE for process control (OPC), XML files, or even an HTTP browser. Supporting both internal and external data sources allows the operator displays to be used for a wider range of functions.

The ability to activate interfaces external to the operator graphics is a powerful feature. Examples of this include calling up menus in-context related to applications external to the graphics—for example, bringing up the menu to calibrate a device from the graphics system, as illustrated in Figure 8-7.

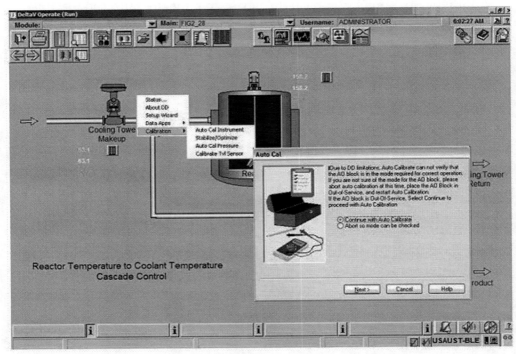

Figure 8-7. External Interfaces – Device Calibration

References

1.   ISA-5.5-1985 - *Graphic Symbols for Process Displays*. Research
     Triangle Park: ISA, 1985 (ISBN: 0876649355).

2.   Treisman, A. "Search, Similarity, and Integration of Features
     Between and Within Dimensions." *Journal of Experimental
     Psychology: Human Perception & Performance* Vol.17, No. 3 (1991),
     652–676.

3.   EEMUA 191 *Alarm Systems – A Guide to Design, Management, and
     Procurement, 2nd edition*, Engineered Equipment and Materials
     Users Association, 2007 (ISBN: 0-85931-155-4).

4.   NAMUR *NA102: Alarm Management*. Leverkusen: User Association
     of Process Control Technology in Chemical and Pharmaceutical
     Industries, 2003.

5.   ANSI/ISA-18.2-2009 - *Management of Alarm Systems for the Process
     Industries*. Research Triangle Park: ISA, 2009 (ISBN: 978-1-936007-
     19-6).

9
Process
Characterization

The concepts that will be addressed in this chapter on process characterization set a foundation that is needed to understand and work with a process control system. It is important that the material presented in this chapter be understood since later chapters build on many of the ideas that are presented.

Process characterization is the identification of the dynamic responses of a process to changes in its inputs. If a control system is to perform well, its design must be based on a good understanding of the process. Therefore, when you are working with control systems, it is important to be able to identify and knowledgeably discuss the dynamic responses of the process to input changes.

9.1 Process Structure

When you are working with the controls associated with a piece of equipment, it is important to understand the relationship of that piece of equipment to other equipment in the plant. The reason for considering how a piece of equipment fits into the rest of the plant is that other process areas may impact its operation—and it may impact theirs. For example, the primary purpose of a gas plant (part of a petroleum refinery) may be to produce gasoline. However, a number of other products such as propane and ethane may also be produced. To manufacture these different products, the feed stream to the plant is processed through a series of distillation columns to separate the various components, as is illustrated in Figure 9-1.

As an aid in studying plant operation, it is often helpful to examine the relative flow rates through different sections of the plant. When such an analysis is done, the primary feed stream to the plant is designated as 1.0 times some flow rate. Based on a primary feed stream flow rate of 1.0, it is possible to easily examine the relative flow rates within the process, independent of the actual feed stream flow rate. For instance, after the primary feed stream is processed through the De-Ethanizer, about 12% of the primary feed stream is separated out from the overhead stream as ethane, and the remaining 88% of the primary feed stream is processed using the De-propanizer column. The propane separated from the overhead flow represents 20% of the primary feed stream and the remaining 68% is further processed by the De-Butanizer. The material separated by the De-Butanizer then becomes the input to the De-Iso-Debutanizer column.

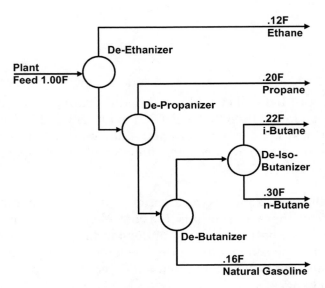

Figure 9-1. Example – Gas Plant Process Flow Diagram

The information provided in a process flow diagram is often sufficient to do such an analysis. By looking at the overall process flow in this manner, it is possible to get an idea of the relative flow rates through the process and the impact that the operation of upstream equipment may have on the operation of downstream equipment. For example, if an operator should report a problem maintaining the product specifications at the De-Iso-Butanizer, then in exploring this problem, it is essential to understand how this unit relates to the rest of the plant. For example, variations in the De-Ethanizer operation can impact the feed stream flow rate and composition to the De-Propanizer and thus could make it difficult to achieve the target Propane production. The controls for the De-Iso-Butanizer may be functioning as designed and the product variation may be caused by operation problems in upstream equipment such as the De-Ethanizer.

This gas plant example illustrates how the operation in one area of a plant may impact the operation in other areas of the plant. Many other examples that apply to the process industry could be mentioned. For example, the steam used in most plants is supplied by boilers located in the power-house area. If the powerhouse cannot respond quickly enough to changes in steam demand, then the pressure of steam in the supply headers (main supply lines) may vary and thus impact the operation of all processes in the plant that use steam. Common services such as steam generation are potential sources of process variation and thus are always things to look out for when troubleshooting a control system.

When you have achieved an understanding of how the various process areas of the plant are interconnected, it is possible to focus in an informed way on an area of interest. In the gas plant example, this might be the De-Iso-Butanizer area shown in Figure 9-2. Based on process flow diagrams and Piping and Instrumentation Diagrams, it is possible to understand the individual significance of the measurements and controls that have been provided to assist the operator in running a piece of equipment. When troubleshooting a problem in measurement or control, your focus may shift to a particular measurement or the control associated with a piece of equipment.

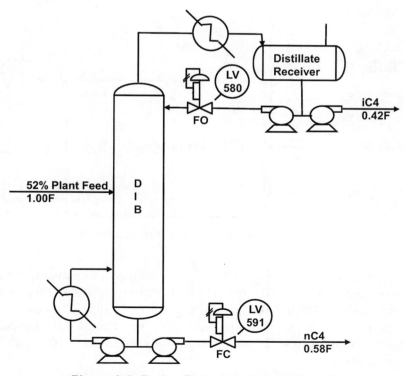

Figure 9-2. De-Iso-Butanizer Process Detail

9.2   Process Definition

As mentioned, when you are working in a process area of a plant, it is often helpful to focus on an individual piece of equipment of interest and its associated measurements and controls. For example, in a reactor area, a heat exchanger can be used to maintain reactor temperature. Other pieces of equipment are required for reactor operation, such as a feed pump and regulating valve. From an analysis standpoint, groupings of equipment in an area of interest, such as the heat exchanger, or the pump and regulating valve in this example, will be referred to as a "process," for example, a flow process, a heat exchanger process.

> **Process** – Specific equipment configuration (in a manufacturing plant) which acts upon inputs to produce outputs.

The equipment that makes up a "process" is not predefined, from an analysis standpoint. For example, an engineer may select a segment of a process area, draw a figurative line around it, and say, "This is the process of interest." Once the equipment that makes up a "process" has been defined, then it is necessary to identify process inputs and outputs. This approach of conceptually breaking a plant process area into individual processes is fundamental to process characterization and control design. In general, a process defined in this manner may be represented as shown in Figure 9-3.

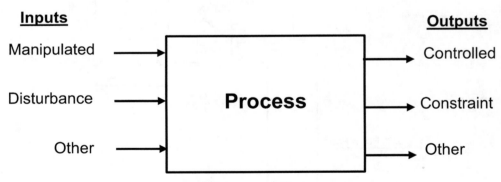

Figure 9-3. Process Representation

As is illustrated in Figure 9-2, a process may have multiple inputs and multiple outputs. To support process analysis and control system design, inputs may be divided into three groups: manipulated inputs, disturbance inputs, and "other" inputs. Similarly, the process outputs may be divided into three groups: controlled outputs, constraint outputs, and "other" outputs.

A controlled process output is also referred to as a controlled parameter of the process. A controlled parameter is defined in the following manner:

> **Controlled output (controlled parameter)** – Process output that is to be maintained at a desired value by adjustment of process input(s).

The value at which a controlled parameter is to be maintained by the control system is known as the setpoint.

> **Setpoint** – Value at which a controlled parameter is to be maintained by the control system.

The primary purpose of a regulating control system is to maintain a controlled parameter at its setpoint. A process input that is adjusted to maintain the controlled parameter at setpoint is known as a manipulated input or manipulated parameter.

> **Manipulated input (manipulated parameter)** – Process input that is adjusted to maintain the controlled parameter at the setpoint.

Another type of process input is a disturbance input.

> **Disturbance input** – Process input, other than the manipulated input, which affects the controlled parameter.

Both a change to a manipulated input and a change in a disturbance input will cause a controlled parameter to change or be driven away from the setpoint.

Some process outputs fall into the category known as "constraint."

> **Constraint output (constraint parameter)** – Process output that must be maintained within an operating range.

From a control perspective, a constraint parameter has no impact unless its value exceeds a constraint limit. The constraint limit is a value that the constraint parameter must not exceed for proper operation of the process.

> **Constraint limit** – Value that a constraint parameter must not exceed for proper operation of the process.

There may be some inputs to the process that have no impact on controlled or constraint outputs. These inputs may be classified as "Other" and are not of interest from a control standpoint.

> **Other input** – Process input that has no impact on controlled or constraint outputs.

Similarly, a process may have process outputs other than controlled or constraint outputs. These types of outputs may be classified as "Other" and are not of interest from a control standpoint.

> **Other output** – Process output other than controlled or constraint outputs.

Unfortunately, the basic terms used to describe a process, such as "manipulated input" or "controlled parameter," are not used in a consistent manner within the process industry. For example, a published article in a magazine or trade journal may make reference to a "control parameter." However, only after reading the article does it become clear that what the author called a "control parameter" is what we have defined as a manipulated input (manipulated parameter). This loose use of terminology can lead to confusion, which can be a barrier to communication. The terminology as defined in this section will be used consistently within this book.

The terminology associated with a process (as we have defined "process" in this chapter) may at first seem abstract. The use of this terminology is best illustrated by considering a practical example: If the analysis of a process is focused on equipment consisting of a piece of pipe that has a valve and a flow transmitter, then this equipment configuration might be called a flow process. The valve position and the pressure upstream of the valve may be identified as the process inputs. The flow rate and pressure downstream of the valve are process outputs. Since the flow rate is measured, this could be a controlled parameter of the process. The valve position would be adjusted to maintain the flow rate at the setpoint. Thus, the valve position would be the manipulated parameter. Since the pressure upstream of the valve will impact the flow rate, then the upstream pressure would be a disturbance input. Since, in this example, there are no limits on the downstream pressure, the downstream pressure would be classified as an "Other" output. A representation of this flow process is shown in Figure 9-4.

In considering a process, one interesting observation is that if there are no disturbance inputs to the process, then by simply adjusting a manipulated parameter, it should be possible to move a controlled parameter to its setpoint. The controlled parameter would remain at setpoint and there

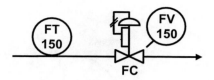

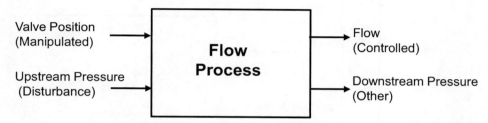

Figure 9-4. Example – Flow Process

would be no need for a control system to automatically adjust the manipulated parameter. Fortunately for the manufacturers of process control systems, most processes have many disturbance inputs. Therefore, a means of
automatic control is needed to maintain the controlled parameter at setpoint by adjusting the manipulated input to compensate for changes in the
disturbance inputs. Both changes in disturbance and manipulated inputs
will impact the controlled parameter as is illustrated in Figure 9-5.

A step change in the valve position—the manipulated input to the flow
process—results in the flow rate—the controlled parameter—changing in
value, as would be expected.

> **Step Change** - A sudden change from one steady-state
> value to another.

Similarly, a step change in the upstream pressure—a disturbance input—
will cause a change in flow. To maintain the flow rate at setpoint, an automatic means is needed to adjust the manipulated input to compensate for
changes in the disturbance input.

A more complex example from the pulp and paper industry can be used to
further illustrate the application of process terminology. Within a pulp
and paper plant, there is a process area called recausticizing. In this area, a
piece of equipment called a lime mud filter is used to remove water from a
slurry stream that contains calcium carbonate.

> **Slurry** - A suspension of insoluble particles in a liquid.

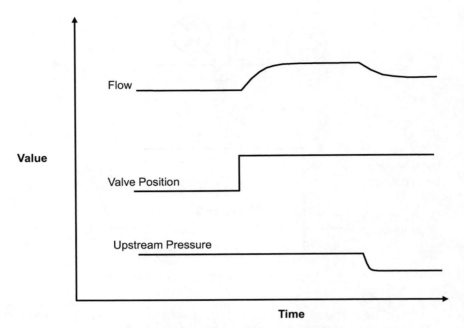

Figure 9-5. Flow Process – Response to Input Change

A sketch of the lime mud filter and how it is typically instrumented is shown in Figure 9-6.

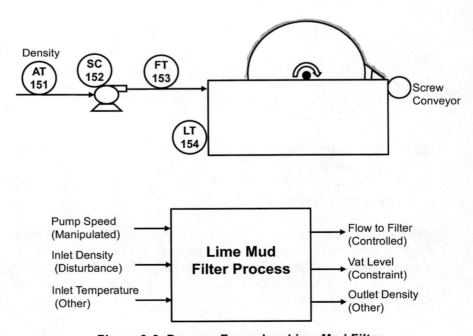

Figure 9-6. Process Example – Lime Mud Filter

The slurry feed to the lime mud filter may contain 30% calcium carbonate by weight, with the remaining material primarily being water. The slurry flow rate into the filter is regulated using a variable speed pump (SC152 in the figure). The lime mud filter consists of a vat that contains a fabric-covered drum. A vacuum pump (not shown in this figure) is used to pull a vacuum inside the drum. As the drum rotates in the vat, the material clings to the fabric and much of the water is removed by the vacuum pump. The solid material that remains is scraped off by a fixed blade as the drum rotates, and is removed by a screw conveyor. In this way, much of the water in the slurry is removed. The flow of slurry is measured using a magnetic flowmeter, FT153. The level in the vat, LT154, is measured using a pressure transmitter with a purge stream. The slurry stream density, AT151, is measured using a nuclear density analyzer.

Applying process terminology to the lime mud filter process, the manipulated input is the pump speed. The volumetric slurry flow rate to the filter is the controlled parameter. Since the inlet density impacts how easily the slurry may be pumped (the higher the density, the greater the resistance to flow), any change in the density measurement is a disturbance input. Vat level is a constraint parameter since the level must be maintained below a maximum level to avoid material overflowing the vat. The inlet temperature and outlet density are unmeasured and do not impact the flow and thus are classified as other. As the pump speed—the manipulated input—increases by a step, the slurry flow rate to the mud filter—the controlled parameter—changes as illustrated in Figure 9-7. As a result of the flow increase, the level starts to increase in the vat. As the level increases, the amount of slurry in contact with the drum fabric also increases until the quantity of solid material that is removed by the rotating drum matches the quantity of solid material in the slurry flow to the vat. As illustrated, the level increases and stabilizes at a new value. If the inlet density increases, then the quantity of material removed by the fabric may actually go up and the level may go down for the same flow rate.

Under normal operating conditions, the speed of the pump may be adjusted to achieve the setpoint of the flow control parameter. However, over time, less material is removed by the filter as the fabric starts to plug. When this happens, the vat level will rise. When the level reaches the constraint limit, then the pump speed must be reduced to avoid exceeding the constraint limit. Under this condition, the controlled parameter, slurry flow, would no longer be maintained at setpoint.

As will be covered in later chapters, the automation of constraint limit handling is done through override control. This is an example of a process that includes a constraint limit that could be easily addressed using override control.

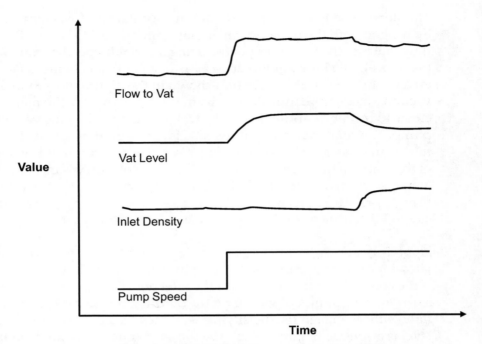

Figure 9-7. Lime Mud Filter – Response to Input Change

9.3   Pure Gain Process

The inputs and outputs associated with a process (and defined in the previous section) have a large impact on the techniques that may be used to control the process. Also, the manner in which process outputs respond dynamically to changing process inputs has an impact on control system design and commissioning.

> **Commissioning** - The work required to bring a system to the point where it can be used.

The dynamic process response can be observed by applying a step change to a process input and observing the changes in the process outputs. During such a test, it is assumed that only one input is changed; all the other inputs are maintained at a constant value and thus have no impact on the results of the test. However, when step testing is done in an operating plant, it is often not possible to maintain all the other process inputs constant. To accurately establish the dynamic process response to a change in process input, it is often necessary to repeat each step test multiple times and to average the results.

If a step change is introduced into a process input and the output immediately duplicates the input change, except for a gain change, then the process is characterized as a pure gain process. The step response of a pure gain process is shown in Figure 9-8.

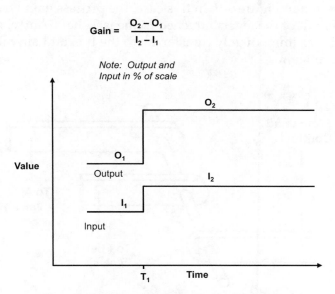

Figure 9-8. Pure Gain Process Response

When working with a pure gain process, only the process gain must be known to fully describe the dynamic response of the process.

> **Process gain** – For a step change in a process input, the process gain is defined as the change in process output divided by the change in process input. The change in output should be determined after the process has fully responded to the input change.

Implicit in this definition is that the changes in output and input are expressed in terms of percent of scale. In many control systems, the process inputs and outputs are often expressed in engineering units, so the engineering unit values must be converted to percent to calculate the process gain.

A pure gain process is an example of a process that is self-regulating. As demonstrated by a pure gain process, the process outputs seek a new steady-state value after a change in process input.

> **Self-regulating** - The ability or tendency of the process to maintain internal equilibrium.

Many of the best examples of pure gain processes are mechanical systems. For example, on older boilers and heaters, control of the fuel and air required to meet steam or heat demand was achieved by using a jackshaft as illustrated in Figure 9-9. In this case, the process gain was determined by the length of the levers attached to the jackshaft. Any change in demand was immediately transferred to the fuel and air valves with only a change in gain.

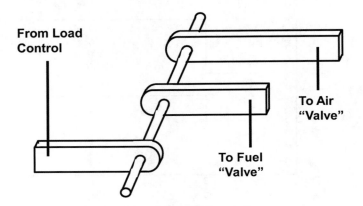

Figure 9-9. Pure Gain Process Example

The response of a liquid process such as a variable speed drive on a pump is extremely fast, so the change in liquid flow for a step change in the input to the variable speed drive closely approximates the response of a pure gain process.

9.4   Pure Delay Process

A step change in the input to a process may have no impact on the process output for a period of time, and when the output does change, it follows the input change except for a delay and a gain change. Such process is described as a pure delay or pure deadtime process. The response of a pure delay process is shown in Figure 9-10.

The dynamic response of a pure delay or pure deadtime process is fully captured by identifying the process gain and deadtime. Remember that the process gain is the percent change in output divided by the percent change in input after the process has fully responded to the change in input. The process deadtime is the time measured from the start of process input step change until the process output value begins to change.

> **Process deadtime** – For a step change in the process input, process deadtime is defined as the time measured from the

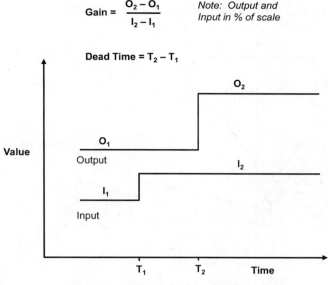

Figure 9-10. Pure Delay Process

input change until when the first effects of the change are detected in the process output.

Process deadtime is also commonly referred to as process delay. The delay or deadtime associated with the process response may be due to measurement processing time or it may be caused by transport delay within the process equipment. In some cases, for example, the processing time associated with a sampling analyzer can introduce significant delay into the observed process response to a step change in a process input. Thus, an important consideration in selecting an analyzer is the measurement delay introduced by the signal processing or by a sampling system. The time for material to move through the process, known as transport delay, may contribute to or be the main source of process deadtime. Selection of transmitter location within the process can often impact the transport delay.

The movement of solid material from a silo using a conveyor belt as shown Figure 9-11 is an example of where the process equipment introduces significant process transport delay. In this example, the rate of extraction of material from the silo is determined by the speed of a screw feeder. The extracted material is transported by the conveyor belt, and a weigh scale near the end of the conveyor is used to measure the rate of flow of material. In this case, a change in screw speed will not be seen until the material has been transported to the weigh scale. The process delay is determined by the conveyor belt speed and the distance from the silo to the weigh scale. The measurement made by the weigh scale, WT158, exactly represents the weight of material extracted from the silo but is delayed by the transport time associated with the conveyor belt.

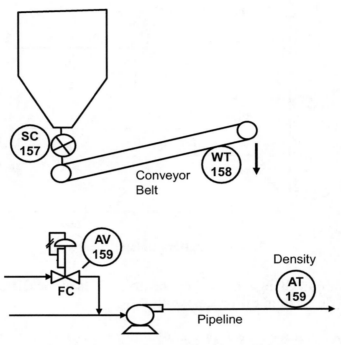

Figure 9-11. Example – Pure Delay Process

The pipeline example shown in Figure 9-11 also shows how transport delay may be introduced into a liquid stream. In this example, a liquid stream of higher concentration is added to a stream of lower concentration. By adjusting the flow rate of the higher concentration stream, it is possible to change the concentration of the combined stream. The impeller of the pump helps to blend these materials to a uniform concentration after the pump. At some point further down the pipeline, the combined liquid concentration is measured using analyzer AT159. In this example, a change in the high concentration flow would not be seen by the analyzer until the combined liquid has been transported down the pipeline to the analyzer. Such an arrangement has been used in the pulp and paper industry to prevent the density in the feed to an evaporator from falling below a minimum value. In one instance, the transport delay was approximately 45 seconds because of the long distance from the pump to the analyzer. As will be addressed in later chapters, such process deadtime can negatively impact control performance. (Note that in some cases process deadtime can be reduced by careful location of the transmitter).

The measurement and control of paper moisture content on a paper machine is another example of where significant process deadtime is introduced by transport delay. A paper machine (Figure 9-12) may be quite long, perhaps 500 feet.

Scanner for Paper Properties   **Final Product**

Figure 9-12. Paper Machine Example

An adjustment made at the front end can take a long time to be seen in the final product since the paper sheet must travel the length of the machine before the final product moisture content can be measured, as illustrated in Figure 9-13. The delay is so large in this case that it makes it difficult for traditional control tools, such as the PID algorithm, to be used for moisture content control. Because of the importance of maintaining tight control of moisture content, companies that supply analyzers for paper properties have developed special tools to improve the control of processes (such as this one) with significant deadtime. Also, deadtime compensation techniques such as the Smith Predictor and model predictive control are provided in some distributed control systems to address the control of processes that are characterized by significant transport delay. [1] [2]

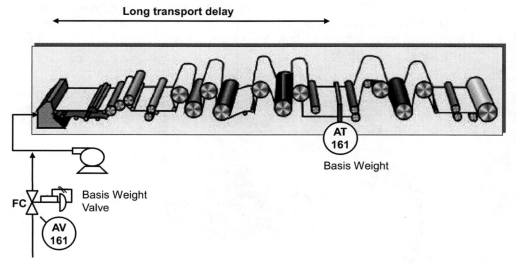

Figure 9-13. Example – Paper Machine Basis Weight

9.5   Pure Lag Process

In some cases, a process output will immediately begin to respond to a step change in a process input—that is, there is no delay. However, the rate of change in the process output is proportional to its current value and the final value the output will achieve for that change in input. A process that exhibits such a response is described as a "pure lag" or "first-order" process (Figure 9-14).

The dynamic response of a pure lag process is fully captured by identifying the process gain and the process time constant. As we have seen, the gain is the (percent) change in output divided by the (percent) change in input after the process has fully responded to a step change in the input. The process time constant is the time measured from the first detectable change in process output until it reaches 63% of its final change in value. (Note that this is *not* the same thing as 63% of its final value.)

> **Process time constant** – For a step change in process input, the time measured from the first detectable change in output until it reaches 63% of its final change in value.

An example of a pure lag process is one consisting of a tank in which the flow rate out of the tank is determined by the tank level and a restriction in the outlet pipe, as illustrated in Figure 9-15. The flow rate into the tank is controlled by a regulating valve. In this case, the controlled parameter is the tank level and the manipulated parameter is the position of the regulating valve. The rate of flow out of the tank changes with the level in the

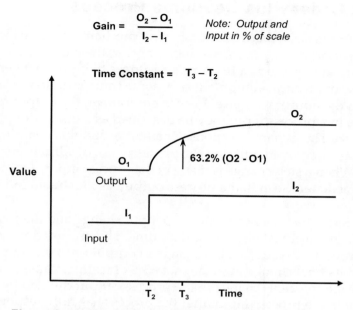

Figure 9-14. Pure Lag (First-Order) Process Response

tank. The level in the tank will settle at a value at which the outlet flow rate matches the inlet flow rate. When a step change is made in the inlet flow rate, the level will immediately begin to change at a rate proportional to the current and final level. Thus, a pure lag response will be seen in the level for a step change in the inlet flow rate.

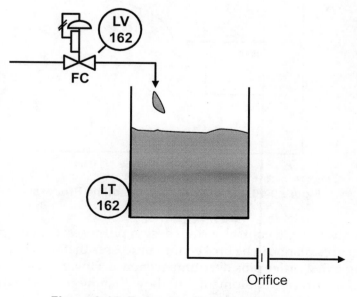

Figure 9-15. Example – Pure Lag Process

9.6   First Order Plus Deadtime Process

Many of the basic thermal and physical mechanisms reflected in the design of process equipment—heat transfer, the blending of gas or liquids—tend to provide a first-order response to input changes. Also, because of physical equipment size, some transportation delay is often seen in the output response. In such equipment, the output response to a change in a process input may be described as a first order plus deadtime response. The dynamic response of a first order plus deadtime process is fully captured by identifying the process gain, deadtime, and time constant. When a step change is made in the process input, a response will not be immediately seen in the process output, as illustrated in Figure 9-16.

To review: The time between a step input change and the first detectable change in output is the process deadtime. The process time constant is the time from the first detectable change in output until the output reaches 63% of its final change in value (not 63% of its final value). The process gain is the (percent) change in the process output divided by the (percent) change in the process input after the process has fully responded.

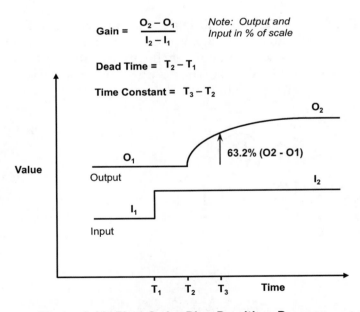

Figure 9-16. First Order Plus Deadtime Process

For analytical and control system design purposes, the equipment that makes up a plant can be broken into processes that are characterized as having a first order plus deadtime response. However, a close examination of the process equipment will show that the process output response may be the end result of many mechanisms working together and interacting: heat transfer, momentum, inertia, air capacitance in an actuator, and so on. The combined effect of these different mechanisms constitutes a higher order process response, but one that often closely (and, for most purposes, adequately) approximates a first order plus deadtime response. One example of a process that may be characterized as having a first order plus deadtime response is a heater, as illustrated in Figure 9-17.

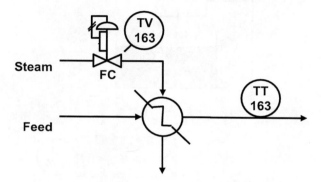

Figure 9-17. Heater Example – First Order Plus Deadtime Process

When a step change is made in a heater steam inlet valve position, a delay will be observed in the heater outlet temperature response. It takes some time for the feed stream flowing through the heater tube bundle to reach the temperature transmitter. Once the temperature starts to change, it exhibits a first-order response. This is because of the mass of metal that makes up the heater and the heat transfer mechanism between the liquid feed inside the tube bundle and the steam that is outside the tube bundle. The combined impact of the transport delay and heat transfer provides a first order plus deadtime response.

Multiple mechanisms working together within a process can sometimes combine to provide what may be approximated as a first order plus deadtime response, even where no transport or measurement delay is involved. Figure 9-18 illustrates how a process that is made up of three lags in series (a third-order system) can be approximated as a first order plus deadtime process.

In this example, three pure lag processes (often simply called *lag processes*) are combined in a series to create a third-order process. The manipulated input is the inlet flow to the first lag process. The output of the first lag process becomes the input of the second lag process and its output becomes the input of the third lag process. The level in the third lag process is the controlled parameter of the combined third-order process. As

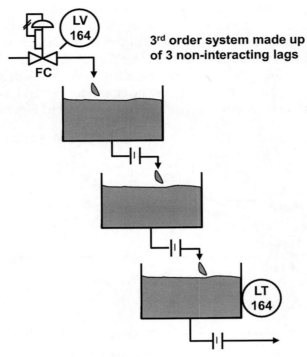

Figure 9-18. Addressing Higher Order Systems

previously discussed, a change in the output of a lag process begins immediately when the process input changes. However, as shown in Figure 9-
19, the combined impact of the three lag processes in series gives a level
output response that closely approximates the response of a first order
plus deadtime process.

9.7   Integrating Process

In some cases, a step change in a process input will result in the process
output changing continuously. A process that exhibits such a response is
known as an integrating process. In everyday life, there are many examples of integrating processes. For example, the response of bathtub level to
a change in water input flow is that of an integrating process. If the valve
regulating water flow is opened, then if the inlet flow exceeds the flow out
of the tub, the level in will increase continuously or at least until the tub
starts to overflow. In the process industry, there are similar examples of
integrating processes such as level control of storage tanks and pressure in
a gas or steam header.

When an input to an integrating process changes, the output may not
begin to respond for a period of time because of measurement or transport

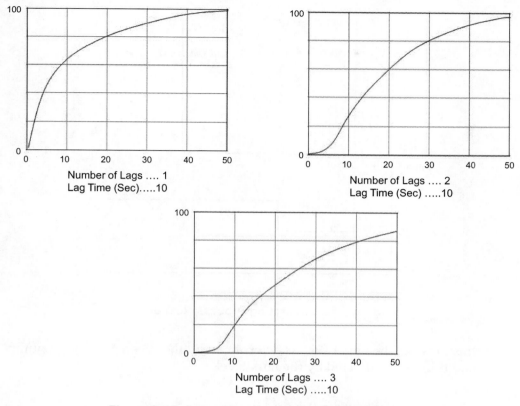

Figure 9-19. Approximating Higher Order Systems

delay. Thus, one parameter that may be used to characterize an integrating process is the process deadtime. Once the process output begins to change in response to a change in input, then the observed rate of change in the output is used to characterize an integrating process. Specifically, the rate of change in the output per percent change in input per second is defined as the integrating gain. Thus, the response of an integrating process in general is described by the process integrating gain and deadtime but in many cases, the deadtime is zero.

> **Integrating gain** – Rate of change in the output of an integrating process per percent change in input per second.

The response of an integrating process and the calculation of the integrating gain are shown in Figure 9-20.

Integrating processes are also described as non-self-regulating processes. An example of an integrating process is a tank where the flow into the tank is the manipulated input, the tank level is the controlled output, and the discharge from the tank is a gear pump as illustrated in Figure 9-21. In

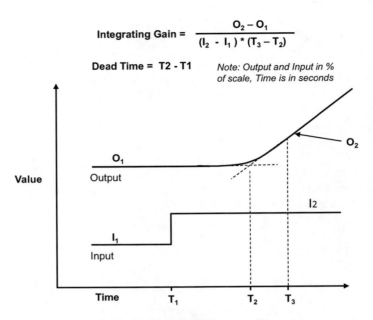

$$\text{Integrating Gain} = \frac{O_2 - O_1}{(I_2 - I_1) * (T_3 - T_2)}$$

Dead Time = T2 - T1   *Note: Output and Input in %
of scale, Time is in seconds*

Figure 9-20. Integrating Process Response

this process, the outlet flow is determined by the speed of the gear pump and thus is not affected by the tank level.

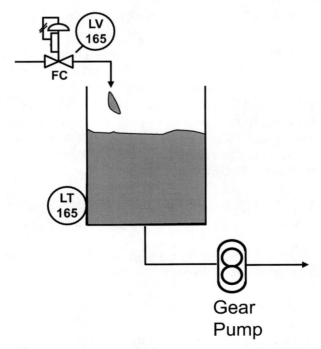

Figure 9-21. Example – Integrating (Non-self-regulating) Process

In this tank level control example, if the inlet flow doesn't match the outlet flow, then the level will constantly change until the tank overflows or goes dry. Also, what this example and the previous tank level example demonstrate is that tank level response is determined by the process design.

9.8   Inverse Response Process

A few processes exhibit a behavior where for a step change in input, the output initially changes in a direction that is opposite to the final direction it will have taken when the process has fully responded to the input change. Such a process is shown in Figure 9-22 and is characterized as exhibiting an inverse response.

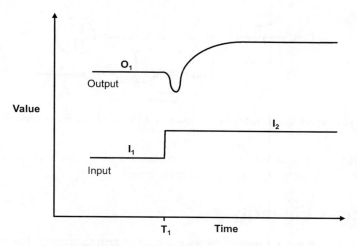

Figure 9-22. Inverse Response Process

Often an inverse response process will only exhibit an inverse response to a step change in one direction of change but not in the opposite direction. The reason for this behavior may be understood when the source of the inverse response is better understood. For example, the level in the bottom of the distillation column illustrated in Figure 9-23 may show an inverse response to a step change increase in heat input.

As a result of a large step change increase in heat input to the column, a lot of bubbling and froth may temporarily result in a false level indication. However, long term, the increased energy input will boil away more material at a faster rate and the actual level will drop at a faster rate than previously. An inverse response has been observed on an increase in energy input. On a decrease in energy input, the same disruption of the level measurement will not be observed. A similar response may be seen in boiler drum level control and other applications. The size and direction

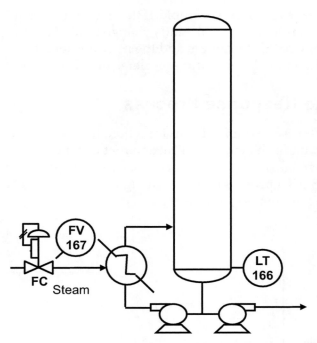

Figure 9-23. Example – Process with Inverse Response

of an input change will often determine whether an inverse response is observed for a change in the process input.

9.9   Process Linearity

Where continuous processing is used in manufacturing, the processes in many areas of the plant will normally operate over a narrow range of operating conditions. However, batch processing often requires that the processes operate over a wide range of operating conditions. As has been mentioned, to characterize a process, a small step change is introduced at normal operating conditions and the results are observed. If such tests are repeated over a narrow operating range then the same results should be observed each time. For example, a consistent process gain, time constant, and deadtime should be identified for a first order plus deadtime process. When the process dynamically behaves the same with the same change in input, the process is described as linear, as shown in Figure 9-24.

Most processes will respond in a linear manner over a narrow operating range. However, over a wider range of operating conditions, the dynamic response that characterizes the process may not be consistent. If the process dynamic response depends on the amplitude of the change in process input then it is said to be non-linear. This change in process response over a wide operating range is typical of nearly all processes and may be caused by the manipulated parameter reaching its upper or lower operating range or the installed characteristics of a final control element such as a valve is non-linear. A process is said to be linear if it meets the criteria illustrated in Figure 9-24.

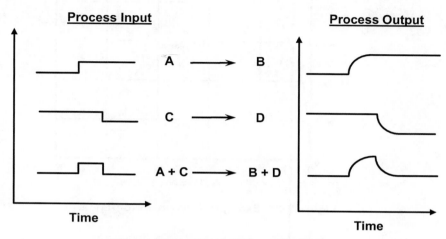

Figure 9-24. Linear Process Criteria

As shown in Figure 9-24, if an input change A provides a response B and an input change C provides a response D, then if the process is linear, the combined input A+C will provide an output response B+D. This simple test may be used to determine whether a process is linear or non-linear. For example, is a pure dead time process linear? Using the above criteria, it becomes clear that a pure dead time process is linear. How about an inverse process? Would it pass this test? Probably not, because the response is sensitive to the direction and magnitude of the step change.

The importance of whether a process is linear over its normal operating range will become clearer in later chapters that address control system commissioning. The setup of the parameters that determine control response, known as *controller tuning*, is based on the identified process characteristics. Thus, the process characteristics used in controller tuning should be those determined in the normal operating region.

The *installed characteristic* of a valve (Chapter 5) can impact the region of operation over which a process provides a linear response. For example, the relationship of fuel gas flow rate to valve position is shown in Figure 9-25 for a valve that is used to control the exit gas temperature of a dryer.

As is illustrated in Figure 9-25, the process gain changes as a function of the valve stem position. For example, a 5% increase in valve stem position

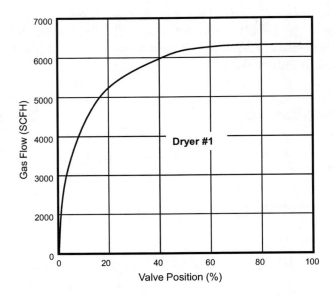

Figure 9-25. Example – Process Non-linearity

when it is operating at a 10% valve stem position might result in an increase in fuel gas flow of 1000 SCFH (standard cubic feet per hour), whereas the same 5% increase in valve stem position at a 40% valve stem position might only increase the fuel gas flow 100 SCFH. The impact of different gas flow changes for the same change in valve stem position would result in different changes in energy input to the dryer. The process is nonlinear due to the installed characteristic of the valve. In this case, the process gain changes with stem position.

This in an illustration of how the valve characteristic selected during the mechanical design phase of a project can impact process response and control performance. Once the field devices have been installed, it may be too costly to modify or replace a final control element.

9.10 Workshop Exercises – Introduction

Workshop exercises provided in this and later chapters are designed to give the reader an opportunity to gain practical experience. The dynamic process simulations used for these workshops provide an environment that closely resembles that of an operating plant.

In a modern digital control system, the measurement, calculation, and control functions performed by the system are defined using a set of standard function blocks. During system configuration, an engineer selects an appropriate combination of blocks to perform the measurement, calculation, and control functions required in the plant. Information flow between these blocks is defined by configured connections between block inputs and outputs. For example, to create a simple control loop (Figure 9-26), an analog input block AI, controller block PID, and analog output block AO would be selected to access a transmitter measurement value, perform a PID control calculation and then transfer the calculated output to the valve. Standard display components defined for these function blocks would be included in the operator interface to allow the operator to access this control loop.

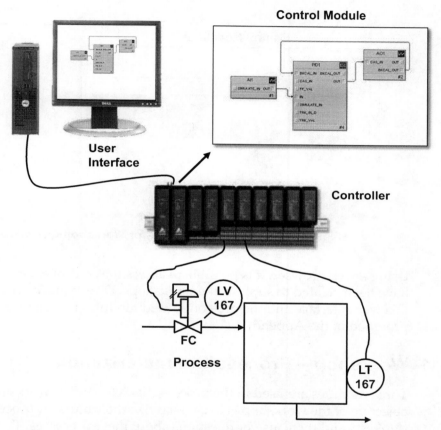

Figure 9-26. Typical Control Loop

To support the workshop exercises, dynamic process simulations have been created that are linked to the function blocks normally used to access field devices, as illustrated in Figure 9-27. Within the accuracy of the process simulation, the parameter values and interactions required when working with the function blocks and operator interface will be the same as in an operating plant.

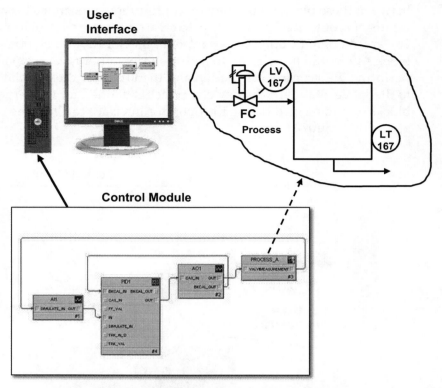

Figure 9-27. Simulated Process for Workshop Exercises

Using a web browser, it is possible to access the control examples that have been created to support the workshops. Directions to access and use this web interface and the pre-configured modules for each workshop are provided in the Appendix.

9.11 Workshop – Process Characterization

Three modules are used in the process characterization workshop. The objective of the workshop is to characterize the simulated processes using step tests and then answer questions about these processes. The trend charts should be used in the workshop to examine the process response to step changes in the process input.

9.11.1 Workshop Directions

Process Characterization-1 Workshop
Using the web interface, open the Process Characterization-1 WorkShop.

Step 1: In the Workshop window, determine the process gain, deadtime and time constant by making a step change, e.g., 40=>50 in the SP parameter of the AO block, and observe the response on the trend. To change the SP parameter, the MODE parameter of the AO block must be set to Auto.

Step 2: Repeat this test, using different SP parameter values, to see if you get the same results.

Question: Based on the observed response, how should the process be classified?

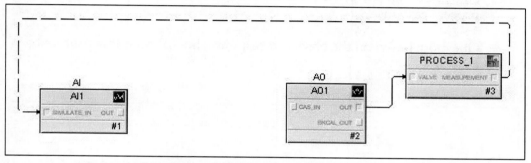

Figure 9-28. Process Characterization-1 WorkShop Module

Process Characterization-2 Workshop
Open the Process Characterization-2 Workshop.

Step 1: Make a step change in the SP of the AO block and observe the process response.

Question: Based on the observed response, how should the process be classified?

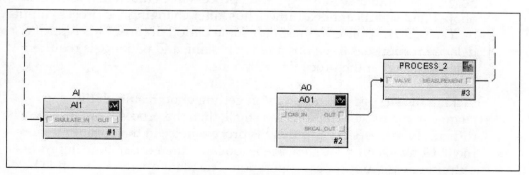

Figure 9-29. Process Characterization-2 Workshop Module

Process Characterization-3 Workshop

Open the Process Characterization-3 Workshop.

Step 1: With the disturbance set to 40, adjust the valve to 50% and observe the response.

Step 2: Once the process output (LEVEL_MEAS) is above 60%, change the AO setpoint to 40 and observe the process output.

Step 3: Change the process disturbance from a value of 40 and observe the process response.

Question: Based on the observed response, how should this process be classified?

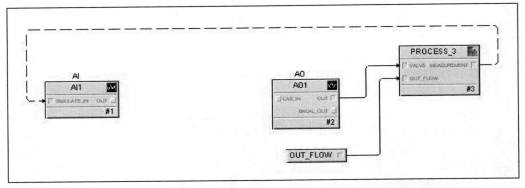

Figure 9-30. Process Characterization-3 WorkShop Module

9.11.2   Workshop Review/Discussion

Step 1: The simulated self-regulating process in CharacterizationWork-Shop-1 had significant deadtime. Thus, after changing the process input by a step, no response was seen for some period of time. This type of delayed response is often observed in a plant, and patience is required when waiting for the process to respond.

Picking the step size is important in getting good results. If the step change in the process input is too small, then the process response may be difficult to see, especially if there is process noise in the output measurement. Conversely, if the step size is too large then the change in process output may disrupt process operation. The plant operator can often be of help in providing guidance on what size of input change can be made without disrupting plant operations.

As shown in this exercise, from the trend of the process response, it was possible to characterize the process. The following results should have been obtained.

> Gain = 1
> Time Constant = 30
> Deadtime = 10

During the commissioning of a control system, it is necessary to characterize the process response to calculate the controller tuning that will be required to provide best process performance. The same procedure used in the workshop would be followed during commissioning, that is, assign a controlled parameter and a manipulated parameter to a trend chart, change the process input by a step, and observe the response. If the process is noisy or has significant disturbance input changes when step testing, then it may be necessary to repeat the test to correctly characterize the process. Based on the identified process characterization, calculate the controller tuning. During a plant startup, this must be done quickly and without disrupting the process operation.

Step 2: An inverse process response was simulated in Characterization-Workshop-2. The response of the process output to a step change in the input should have been observed to be in a different direction than that required to reach the final output value. Addressing an inverse process response is challenging from a control standpoint. The only solution may be to de-tune the control system to avoid its reacting too quickly to input changes, but by keeping input changes small, an inverse response may often be avoided.

Step 3: In CharacterizationWorkShop-3 an integrating process is simulated. The simulation addresses tank level, where the input flow can be adjusted and the outlet flow—the Disturbance input in the simulation—is determined by a downstream process. If the inlet flow and the outlet flow are equal, then the level becomes constant because what's going in is equal to what's going out. As soon as the Disturbance input—outlet flow—is reduced, the level starts to increase because the inlet flow is now greater than the outlet flow.

When working in a plant environment, it is often challenging to do step testing to characterize an integrating process. As with other such testing, the step size must be picked to avoid disrupting the process operation. Often there are disturbance inputs that impact the process response, making it difficult to determine the integrating gain accurately.

References

1.   Blevins, Terrence L. "Modifying the Smith Predictor for an Application Software Package." Paper published in *Advances in Instrumentation, Vol. 34 Part 2*, ISA-79, National Conference and Exhibit, Chicago, Illinois, October, 1979.

2.   Blevins, Terrence L. and Wojsznis, Peter. "Providing Effective Control of Processes with Variable Delay." Paper presented at ISA EXPO 2002, Chicago, Illinois, October, 2002.

10

Control System Objectives

In this chapter, we examine the objectives of a control system. The design of a control system should take into account the desired system performance and the installation and maintenance costs. Having a clear idea of control system objectives is critical to a successful installation. In the days of pneumatic and analog control systems, it was necessary to carefully consider the objectives early in a project since any change in the control system required physical changes to components and wiring. The flexibility of modern digital control systems allows changes in control design to be made quickly, with little or no extra cost. However, such changes often directly impact operator training, display construction, and in some cases, the field devices that will be required to implement the control strategy. So, the flexibility of a digital control system does not reduce the need to clarify the control system objectives early in a project.

The objectives of a process control system generally fall into one of three categories:

- Economic Incentive
- Safety and Environmental Compliance
- Equipment Protection

One objective of a control system is to respond to an economic incentive (cost saving or increased revenue) by an improvement in plant operating efficiency through better utilization of feedstock materials (i.e., higher yields), a reduction in off-specification product, or a reduction in energy requirements. Where there is sufficient demand for a product produced by a plant, then increased profit can be achieved by increasing plant throughput. The economic benefits provided by a new or upgraded control system can be calculated.

The safe and environmentally compliant operation of a plant often requires that critical operating parameters be maintained within their normal range. The control system is designed with safety related functions that, in the event of a process upset, act to avoid the disruption that can be introduced by a safety system shutdown. However, if safe operating conditions are not maintained by the control system the separate safety system takes action. The control logic and field devices that make up a safety system operate independently of the process control system. In this way, the safety system acts as an independent arbitrator of actions taken by the control system, and thus is ultimately responsible for the safety of the plant.

The functions provided by the control system are also essential to (for example) keep stack gases and waste streams within limits set by environmental standards.

The third category of control system objectives is to protect the process equipment. Certain plant operating conditions may result in process equipment damage but do not pose a safety or environmental risk. In these cases, the process control system is often responsible for taking corrective action to avoid equipment damage. For example, the pump that is used to transfer liquid from a tank may run dry if the tank is emptied. This can cause the pump to be damaged. Since this condition does not create an unsafe operating condition, it will not be acted on by the safety system. However, replacing a pump can be costly and production could be disrupted if the pump cannot perform its normal function. In such a case, the control system may automatically shut off the pump to avoid equipment damage.

10.1 Economic Incentive

If a control system can reduce variations in process operation, it is possible to shift the plant to a more efficient point of operation. The economics of plant operation may be impacted by process variation when maximum production and operating efficiency are achieved at a specific operating condition or production is limited by equipment capacity. When production is limited by equipment capacity, then maximum production is achieved when the control system minimizes variations and adjusts operating targets that allow this equipment to be operated closer to their operating constraint. When maximum production is achieved by maintaining specific operating conditions then such a process is referred to as exhibiting a global production maximum. In such cases, any variation from these operating conditions will reduce production.

The generation of steam in a boiler is an example of a process that exhibits a global production maximum at a given throughput. The amount of excess air in the flue gas leaving a boiler has a direct impact on boiler efficiency and the cost of steam generation. When the amount of air delivered to the boiler is greater than that required for complete combustion, the excess air leaving the boiler decreases the boiler efficiency because air is being heated in the combustion process. As the excess air level in the flue gas is decreased, at some point only enough air will be provided to support complete combustion of the fuel. Boiler efficiency will go down if the combustion air level is reduced further, and unburned fuel will exit the boiler in the flue gas.

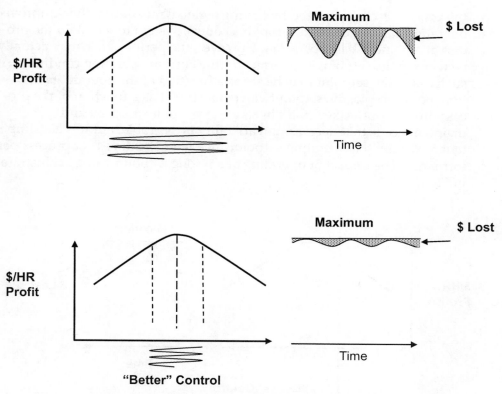

Figure 10-1. Operating at Global Production Maximum

As illustrated in Figure 10-1, for a given firing rate the best operating efficiency is achieved at a specific operating point. Any variation from this operating point will result in decreased operating efficiency (and, it should be noted, possible environmental non-compliance). If the control system can reduce variation in the excess air level, the boiler operating efficiency can be increased. For a given firing rate, the impact of excess air level on boiler efficiency will exhibit a global production maximum.

As described in Chapter 4, a measurement of the O_2 concentration in the flue gas exiting the boiler may be used as an indirect indication of the boiler excess air level. If O_2 control has not been implemented then variations in the excess air level will occur with changes in the process inputs. Examples of such changes are variations in fuel BTU/lb value or a change in the air density because of a change in the outside temperature. The boiler control system must minimize variation in this key process parameter with changes in process inputs. The O_2 setpoint in boiler combustion control is often characterized as a function of the load (demand) on the boiler because the most efficient operating point may change as a function of the firing rate, which changes with changes in load.

If a process is characterized by having a global maximum, then improved operation can be achieved through a control system to maintain the process at a point of best operation. However, the performance may actually be worse if the setpoint is incorrect for the current operating conditions. In such cases, the setpoint may be set as a function of the operating condition. For example, the setpoint valve may be set as a function of the process throughput. A key point in control system installation and maintenance is that decreasing variability is desirable, but it is also important to specify the operating setpoints needed to achieve best process performance. The impact of operating at a wrong setpoint value is illustrated in Figure 10-2.

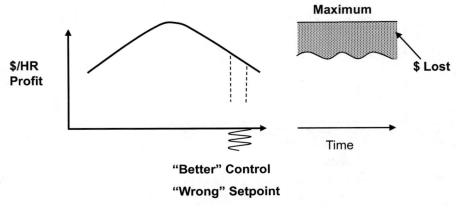

Figure 10-2. Operating Away from the Global Production Maximum

When a plant is designed, the equipment is sized so that at the target production rate, the equipment in each area of the plant is operating at near maximum capacity. Even so, as plant production is increased above the design production rate, one piece of equipment may become the bottleneck. If plant production is not limited by the market, that is, an increase in production can always be sold, then maximum profit may often be achieved by running the plant at maximum throughput. If production can not be increased because of a physical limit in equipment operation, by reducing variation in the process operation, it is often possible to shift the production target to increase plant production without exceeding the physical limit in equipment operation as illustrated Figure 10-3.

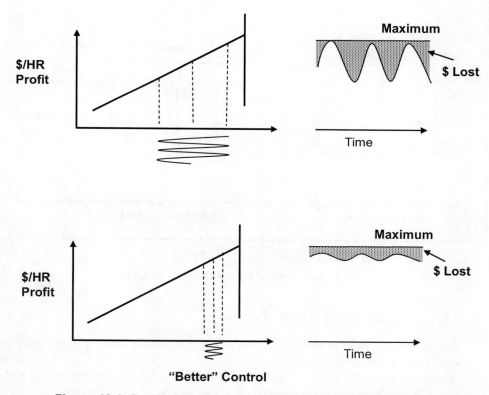

Figure 10-3. Production Maximum at Equipment Physical Limit

The term *production-limited* is often used to describe the condition where the maximum production of a plant is determined by limitations in one or more processes. In such cases, it is important to understand, "Where is the plant bottleneck?" That is, which process area or piece of equipment limits plant production? One way to determine this is to ask the question, "Why can't plant production be increased?" The answers to such a question may help clarify which pieces of equipment limit plant production.

If the control system allows wide variations in plant operation, the target production rate must be set at a value that ensures that the limit of each piece of equipment is never exceeded. By improving the control system to reduce process variation, plant production may be increased without exceeding the limits associated with the process area that is the plant bottleneck, as illustrated in Figure 10-4. This simple concept is often the basis used to justify the cost of upgrading or installing a better control system in a plant.

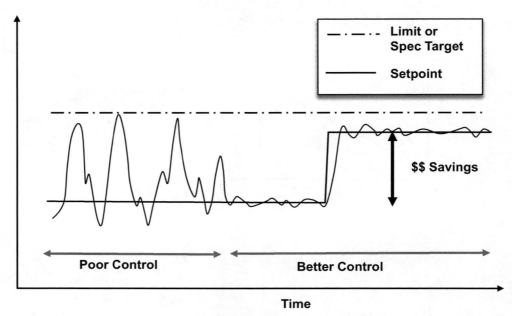

Figure 10-4. Production Maximum at Limit

10.1.1 Ammonia Plant Example

The impact of process variation on plant economics can be illustrated by examining the operation of a typical ammonia plant. The feedstock to an ammonia plant is natural gas and so plants are located in places where there is an abundance of natural gas. There are a few ammonia plants in North America but natural gas is so expensive that it is a challenge for these companies to compete on the world market. There is a strong incentive to operate these plants in the most efficient manner.

The process flow through an ammonia plant is shown in Figure 10-5. In the primary reformer section, the natural gas feedstock (methane, CH_4) is combined with steam (H_2O), and heated to a very high temperature in the presence of a catalyst. As a result of the reaction with the catalyst, this feedstock is converted to hydrogen (H_2) and carbon monoxide (CO). In the secondary reformer, air (approximately 21% O_2, 78% N_2, and some inert gases such as argon) is added and in a combustion process, most of the carbon monoxide combines with the oxygen leaving CO_2, H_2, and N_2. Any unreacted CO is converted to CO_2 in the shift reactor. In the following absorber units, the carbon dioxide is removed, leaving nitrogen and hydrogen. Any unabsorbed CO_2 is then converted to methane (CH_4) in the Methanator.

In the downstream process, the feed stream of H_2, N_2, and traces of methane and inert gases is compressed. In the synthesis loop, this feedstock is passed over the catalyst in the synthesis converter where the hydrogen and nitrogen combine to form ammonia, NH_3. A separator is used to remove ammonia from the synthesis loop. Any accumulation of methane and inert gases in the synthesis loop is removed using a purge stream.

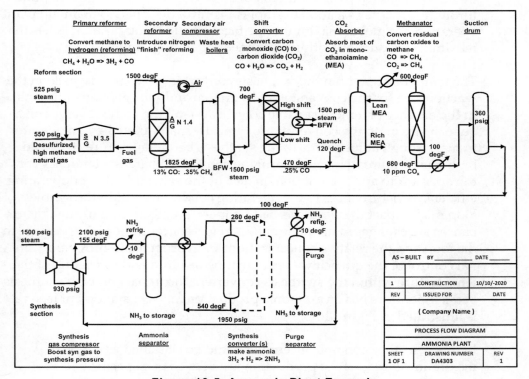

Figure 10-5. Ammonia Plant Example

The control system provided with a new ammonia plant is often designed to address normal day-to-day plant operation. Such a basic control system is quite simple and allows the plant to be quickly commissioned and brought up to design operating conditions. However, once stable plant operation has been achieved, to maximize operating efficiency and production (as is important in North American plants), it is often necessary to invest in added control capability. Before the introduction of distributed control systems, it was necessary to add a computer to the control system to provide this additional control capability. At one time, process control suppliers offered special ammonia plant control packages, which included services to engineer and install computer-based control systems for plant optimization. However, if a plant has a modern distributed control system, the functionality of the control system may be sufficient to implement control designed to increase plant operating efficiency and production.

A significant improvement in ammonia plant operation efficiency and throughput can often be realized through the addition of control in four or five areas of plant operation. This added control directly impacts the operation of critical units within the plant. In most cases, it is possible to achieve an investment payback of only a few months; thus, the controls that may be added to an ammonia plant to improve plant efficiency and throughput serve as a good example of control designed to maintain plant operation at a global maximum in efficiency and throughput at levels that maximize production within equipment operating limits.

The temperature within the synthesis converter has a large impact on the reaction rate for the conversion of hydrogen and nitrogen into ammonia. In the absence of a more capable control system, the manual valves on a quench stream to the converter are typically manipulated by an operator to make rough adjustments to the synthesis bed temperature. To avoid high temperature that could harm the converter, there is a tendency for the operator to apply too much quench at the expense of lower operating efficiency. Disturbances to the operating conditions within the synthesis loop may impact the synthesis bed temperature. Examples of such disturbances are changes in methane and inert gas concentration, the pressure in the loop and the relative concentrations of hydrogen and nitrogen. Manual control of the quench valves may thus result in wide variation of the temperature within the synthesis converter. The impact that the synthesis bed temperature has on ammonia production for a constant plant feed rate is illustrated in Figure 10-6.

In this case, the economic incentive would be classified as a global maximum since greatest production is achieved at a specific bed temperature under normal operating conditions. At a given feed rate, ammonia plant production may be reduced by any reduction in the synthesis bed temperature from its optimal value. There is a strong incentive to install control capability that will automatically regulate the synthesis converter quench valve to maintain the bed temperature at setpoint. This automatic control will immediately compensate for any changes in the operating conditions. The improvement in efficiency that

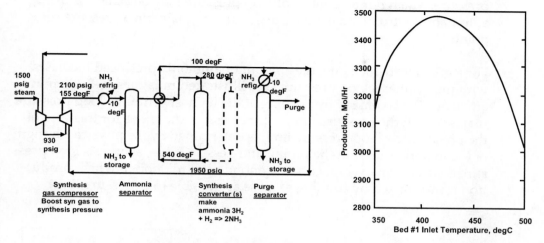

Figure 10-6. Impact of Synthesis Bed Temperature on Plant Production

may be achieved through automatic control of the synthesis bed temperature can be calculated based on reducing the average deviation from the best operating point (optimal value). For example, based on the curve shown in Figure 10-6, an increase in the average operating temperature could increase plant production by 0.4%/°C.

@410°C

$$\frac{\% \text{ production change}}{\text{Change in bed \#1 temperature}} = \frac{\left[\dfrac{(3441 - 3413)Mole\ NH_3}{Hour}\Big/ 3441\dfrac{Mole\ NH_3}{Hour}\right]}{(430 - 410)°C} \bullet 100$$

$$= \frac{0.4\% \text{ production change}}{1 \text{ degree change in bed temperature}}$$

Since an ammonia plant may produce over 1000 tons of ammonia per day, even a small increase in production for a given feed rate can significantly increase the profit generated by the plant.

The operating pressure in the synthesis loop can also impact the efficiency of ammonia plant production. The pressure in the synthesis loop is normally changed by a valve that the operator adjusts to change the purge flow from the synthesis loop. The pressure in the synthesis loop is directly impacted by operating conditions including the plant feed rate, the concentration of methane and inert gases in the feed stream and the rate of conversion in the synthesis loop. When the purge value is manually adjusted by the operator, wide variations may occur in the synthesis loop pressure. Much tighter control of loop pressure may be achieved using automatic control to adjust the purge valve to maintain a pressure setpoint.

For a given feedstock rate, the amount of ammonia produced increases with the synthesis loop pressure as is illustrated in Figure 10-7. Operating efficiency is maximized by running at the highest synthesis loop pressure that is within the pressure limits of the vessels and pipes for this area of the process. The loop pressure limit is determined by the physical strength of the materials used in the piping and vessels. Thus, synthesis loop pressure control is an example of an economic control loop in which production is maximized by running as close as possible to a pressure limit.

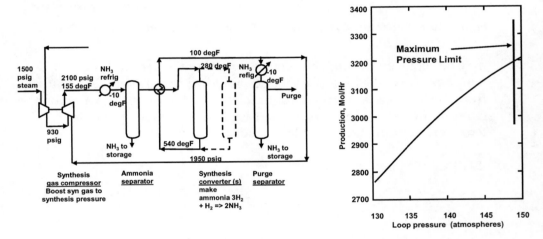

Figure 10-7. Synthesis Loop Pressure Control at a Limit

An average increase in synthesis loop pressure would provide a 0.05%/ Atm increase in production.

@148 *Atm*

$$\frac{\% \ production \ change}{Change \ in \ pressure \ (Atm)} = \frac{\left[\dfrac{(3205.6 - 3142.26)Mole \ \dfrac{NH_3}{Hr}}{3175.24 \ Mole \ \dfrac{NH_3}{Hr}}\right]}{(150 - 146) \ Atm} \bullet 100$$

$$= \frac{0.05\% \ production \ change}{1 \ Atm \ change \ in \ pressure}$$

In the front end of an ammonia plant, the temperature that is maintained in the primary reformer tubes impacts how much of the natural gas feedstock is converted into hydrogen and CO_2. The conversion rate increases with higher temperatures. How close the conversion rate comes to 100% depends on the tube temperature. Any unconverted natural gas is known as methane leakage. As the methane leakage increases, the purge flow in the synthesis loop must increase to maintain synthesis loop pressure at a safe level, reducing ammonia production. For a given feed rate, the impact of methane leakage on ammonia production is shown in Figure 10-8.

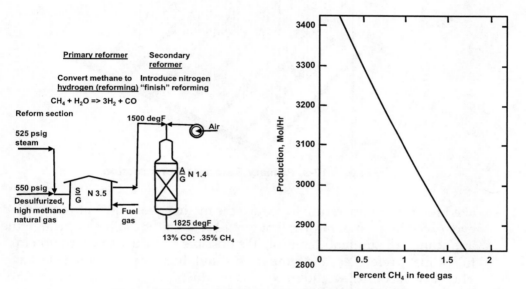

Figure 10-8. Primary Reformer Methane Leakage

The natural gas and steam input streams to the primary reformer flow through tube bundles that are filled with catalyst. Banks of these tube bundles, often referred to as *harps,* are contained inside refractory-lined walls, as illustrated in Figure 10-9. Burners located at the bottom of the primary reformer are used to heat the gases flowing through the tube bundles. Fuel gas valves that are normally manually adjusted by a plant operator determine the energy input to the burners and thus the temperature of the gases flowing through the tube bundles. The gas temperature is approximately 1500°F (815.55°C). However, the temperature will change with the feedstock rate and also with changes in other operating conditions such as the outside air temperature.

There is a maximum temperature which, if exceeded will allow the metal that makes up the tube bundle to distort and eventually rupture. When a tube fails in this manner, it is necessary to shut the plant down until the tube is replaced. Thus, for the sake of efficiency, it is important to operate

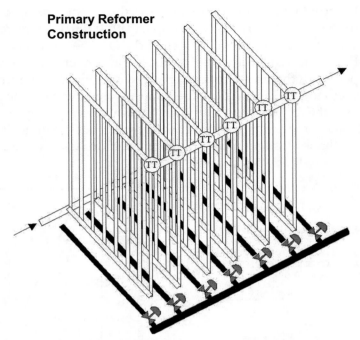

Figure 10-9. Primary Reformer Construction

at a high tube temperature to reduce the methane leakage but without
ever exceeding temperatures that would cause a tube failure. Through the
application of automatic control, it is possible to reduce the variation in
the primary reformer tube temperature and thus reduce methane leakage.
This type of control would maximize production within the temperature
limitations for the primary reformer. For a given feed rate, even a small
reduction in methane leakage can have a large impact on plant production
as shown below.

@1% CH_4 leakage

$$\frac{\% \ production \ change}{Change \ in \ CH_4 \ \%} = \frac{\left[\dfrac{(3303.9 - 2909.1)Mole \ \dfrac{NH_3}{Hr}}{3087 \ Mole \ \dfrac{NH_3}{Hr}}\right]}{(0.5 - 1.5) \ Atm} \bullet 100$$

$$= \frac{12.7\% \ production \ change}{\% \ change \ in \ CH_4 \ Leakage}$$

So, within an ammonia plant there are operation areas where the control
may be classified as global production maximum and others where production is limited by equipment capacity. Hopefully these examples help

demonstrate how, through control, the efficiency of a plant operation may be improved.

10.2 Safety, Environmental Compliance, Equipment Protection

In addition to the economic incentive, the objective of much of the control implemented in a process plant is to maintain the operating conditions needed for safe, environmentally compliant plant operation. In many cases deviation from the normal operating conditions can cause an unsafe condition or have an adverse impact on the environment. Control loops dedicated to this objective are classified as safety or environmental control functions.

Many examples of safety or environmental control functions can be found in a boiler combustion control system. For a "balanced draft" boiler, the air supply to support combustion is provided by a forced draft (FD) fan. Air is mixed with fuel in burners. The proper air/fuel ratio must be maintained to maintain levels of CO emissions below the maximum set by environmental standards.

The hot gases that result from combustion are removed from the boiler using an induced draft fan as illustrated in Figure 10-10.

The pressure of the gases within the boiler combustion chamber can be characterized as an integrating process since the net volume of gases in and out of the boiler must balance to achieve a constant pressure. In this case, the pressure that is maintained in the combustion chamber, known as boiler draft, is important from the standpoint of both safety and equipment protection. If the gas pressure within the combustion chamber goes positive, that is, exceeds the pressure outside the boiler, then hot gases could flow out of observation ports and open doors along the walls of the boiler. These hot gases would pose a safety risk to anyone walking by the boiler. The boiler combustion safety system is designed to shut down the boiler if the gas pressure goes too positive. However, if the boiler draft control is functioning correctly it will take action to always maintain a slight negative draft and thus provide a safe operating environment.

If the boiler draft were to go too negative, that is, the gas pressure inside the combustion chamber were significantly less than the outside air pressure, this could potentially cause the boiler walls to collapse. Such an event could cause millions of dollars in damage to a plant. Thus, boiler draft control served the dual function of providing a negative pressure needed for safe operation and that is within the pressure range the boiler is designed to operate.

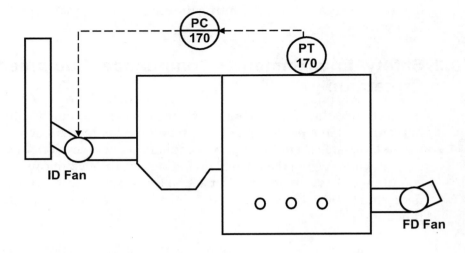

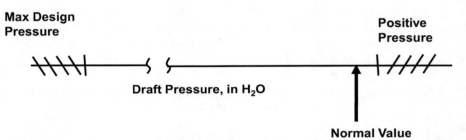

Figure 10-10. Boiler Draft Control

The standpipe level process shown in Figure 10-11 is an equipment protection example. In this example, the input flow rate to the tank may vary with plant operating conditions. If the level in the tank exceeds an overflow level, then the excess material flows into a standpipe and is transferred out of the standpipe using a variable speed pump.

A level transmitter is installed on the standpipe to give a measurement of the level. Based on the level measurement, the variable speed pump is automatically adjusted to maintain a liquid head (maintain the level above zero). By maintaining a liquid head, the pump is protected from running dry and possibly being damaged. This is a case where the objective of the control system is equipment protection and not to maintain the control parameter precisely at setpoint. There is no economic incentive to maintain the level at a specific value. As long as the standpipe level control maintains the level above zero, the control objective has been satisfied.

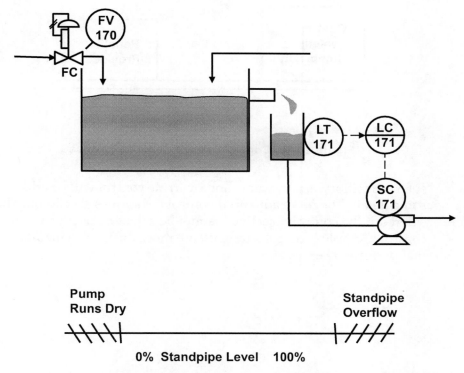

Figure 10-11. Equipment Protection – Standpipe Level Control

As these examples have illustrated, the adequacy of control system performance is determined by how well the system meets the control objectives, that is, economic, safety, environmental and equipment protection.

10.3 Balancing Complexity with Benefits

When designing a process control system the required level of complexity must be justified with measurable benefits to plant operation. This applies whether the objective is to provide basic control functionality for safety or equipment protection or when the objective is to improve the efficiency of plant operations. In all cases, the company or plant must consider whether the complexity and expense associated with a new control system or with simply adding new field devices or more advanced control capabilities are justified by the improvement in efficiency, output, safety, and environmental compliance, as illustrated in Figure 10-12.

A number of different control techniques will be introduced in the following chapters of this book. Examples are used to illustrate how these techniques may be combined to address different application requirement. For example, both basic feedback control as well as more advanced techniques

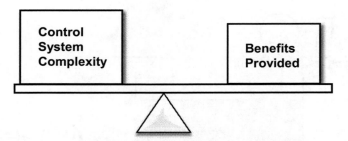

Figure 10-12. Control System Benefits Balance

such as feedforward, cascade, and override control will be addressed in some detail. The more advanced control techniques should only be considered if the control objectives cannot be achieved using basic feedback control. Complex control strategies are more difficult to maintain at optimal performance.

11

Single-Loop Control

Most of the control applications in the process industry are addressed using single-loop control.

> **Single-loop control** – Manual or automatic adjustment of a single manipulated parameter to maintain the controlled parameter at the setpoint.

In this chapter, the basic features of single-loop control and variations in its implementation are explored.

11.1 Manual Control

The most basic means of regulating the operation of a plant is manual control, that is, the plant operator places the control system (or some portion of it) into a manual mode of operation and thus becomes responsible for adjusting valves, setting motor speeds, and so on to run the plant. This may be done because some types of transmitters are not able to provide reliable measurements over the full operating range of a process. For example, during the startup of a plant, it may be necessary to operate the plant at a low throughput rate. The automatic controls cannot be used in many cases because the measurements are not accurate or consistent at the low flow rates that are common under these conditions.

Using manual control, the operator may gradually ramp up the plant throughput until it is up to its normal production rate. Once the normal production rate is achieved, the control system may be switched to automatic operation.

The plant operator may also switch from automatic to manual control during abnormal operating conditions. For example, if work being done on process equipment may impact a control measurement or final control element, then manual control may be required. The natural question then is, "Would manual control ever be selected as the mode of process control for normal plant operation?" It turns out that there are cases where manual control is the most appropriate means of achieving plant operation objectives. For example, if a process has no disturbance inputs or the impact of disturbance inputs is negligible, then the process outputs of interest can be maintained at setpoint by manually adjusting the manipulated process inputs. Once the inputs are set at a value, then the associated process outputs will remain at setpoint without further adjustment to the process inputs.

For this type of process, feedback control provides little or no benefit. Fortunately for the manufacturers of process control systems, most industrial processes have unmeasured disturbance inputs. Therefore, there is a need to implement automatic control to maintain the controlled parameters at setpoint.

For those cases where automatic control is not justified or cannot be implemented, modern digital process control systems provide a manual loader block that is specifically designed to support manual control. Usually this block will be used to implement manual control because of its features to support operator adjustment of the process input. An area where manual control could be used is the feedstock flow out of a storage facility at the front end of a plant. For example, consider the case where pulverized lime flow from a storage silo is manually adjusted by the operator to regulate the lime flow to the process during normal plant operation, as illustrated in Figure 11-1. In this case, the silo is periodically filled from trucks or rail tank cars. Thus, it is important that the operator have an awareness of the level of inventory in the silo. However, the level in the silo need not be maintained at a setpoint. By design, the level of the silo is allowed to vary so that the normal withdrawal of material may be decoupled from the need to refill the silo.

In the case of the lime silo, the level measurement is used by the operator to know whether there is sufficient inventory to continue running the process. Based on the plant production rate and the lime inventory, it will be necessary to schedule truck or rail car delivery to ensure that production will not be disrupted by the silo being empty. Thus, in this case, the silo level is important to plant operation but is not used in the automatic control of the process.

On the P&ID (piping and instrumentation diagram) manual control is indicated by the nomenclature HIC, Hand-Indicator-Control. In many instances, a measurement input is provided for use by the operator in manually adjusting a manipulated input to the process.

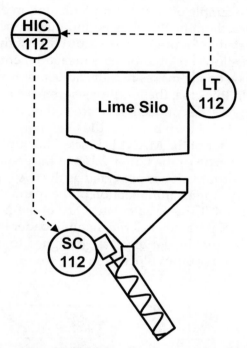

Figure 11-1. Example – Hand Indicator Control

As mentioned, a manual loader block is designed to allow the operator to adjust a manipulated input to the process. Often the output of the manual loader block is used to directly set the control system output to a final control element such as a valve. In the lime silo example, the manual loader block is used to set the target speed of a variable speed drive for a screw feeder (SC in Figure 11-1). By adjusting the screw feeder speed, the operator can directly control the rate at which lime is removed from the silo.

The rate at which lime is used in the process may be determined by the plant throughput or by other plant operating conditions that may not be inputs to the control system. For example, scheduled and unscheduled production changes may be communicated in meeting or by phone. Thus, it may be necessary for the operator to periodically adjust the screw speed. Such an adjustment could be made by the operator directly, by adjusting the input to the variable speed drive. However, the value of using a manual loader block as the operator interface is that another parameter such as the silo level may be shown to assist the operator in making manual adjustments.

11.1.1   Implementation

The manual loader may be implemented as a function block in the control system. An example of such an implementation of a manual loader block, MANLD, based on the Fieldbus Foundation Function Block Specification [1] is illustrated in Figure 11-2. An input (IN) of the manual loader block is provided to allow a related process measurement to be displayed to the operator as he or she makes manual adjustments to the manipulated process input. In this case, the input measurement is provided by an analog input block, AI. The connection between the AI and MANLD block is indicated by the line shown between the OUT parameter of the AI block and the IN parameter of the MANLD block. The output (OUT) of the manual loader block becomes the target value of the analog output block (AO). This target output value is provided as the CAS_IN (Cascade_In) of the AO block. To support initialization of the manual loader output, an output, BKCAL_OUT (Back Calculate_Out) of the AO block is connected to the BKCAL_IN parameter of the manual loader block. The inputs and outputs of the manual loader block are designed to work directly with the analog input and output blocks.

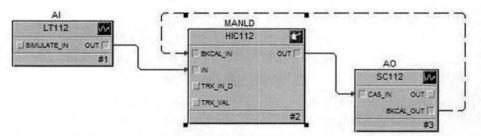

Figure 11-2. Function Block Implementation of Hand Indicator Control

When the operator sets this manual loader output, this value is propagated to the setpoint of the AO block through the connection between these two blocks. The input value provided by the BKCAL_IN allows the manual loader to track the setpoint of the AO block when the AO block is not in Cascade mode. By this means, it is possible to allow bumpless transfer when the mode of the AO block is switched to Cascade mode.

> **Bumpless Transfer** - Method designed to minimize the change in setpoint value on a transition in block mode.

In other words, the OUT parameter of the manual loader must match the AO setpoint before the AO block is allowed to transition to Cascade mode of operation. The use of block mode to control the source of block setpoint and output will be addressed later in this chapter.

As is illustrated by this example, the analog input and output blocks are used in control to access field measurements and to provide outputs to field devices such as regulating valves or variable speed drives. These blocks play an important part in manual and automatic control implementation and for this reason, more details are provided on these blocks in the following sections.

11.1.2   I/O Processing

The method used to process analog inputs and outputs is unique to each control system. For example, the components used in the input and output cards determine the resolution with which measurements may be read and final control elements can be positioned.

> **Resolution** - The ability to detect and faithfully indicate small changes.

Also, controller and bus design influence how often measurement and outputs are updated and thus dictate the control execution rate that is supported by the control system. Understanding the limitations imposed by the basic control system design (i.e., controller and bus design) may be important in achieving the desired control performance.

Measurements of the process operating condition used in control are provided by field transmitters. Traditional field transmitters provide a 4–20 mA signal that is directly proportional to the measurement value. The analog input cards of a control system convert this current signal into a digital representation using an analog-to-digital (A/D) converter. This digital representation is transferred into the controller memory as a 0 to 100% value that may be accessed by an analog input block.

The analog input card typically contains a hardware filter for each input. This hardware filter is used to attenuate any noise contained in the measurement. For example, the current signal may contain 60-Hertz noise because there are AC power lines close to the instrument wires, or process noise introduced by turbulent flow and other variations within the processing may be imposed on the measurement. The hardware filter contained in the analog input card is designed to eliminate any high frequency noise such as variations above 3 hertz from the measurement before it is processed by the A/D converter. An example of the hardware filtering provided on an analog input card is shown in Figure 11-3.

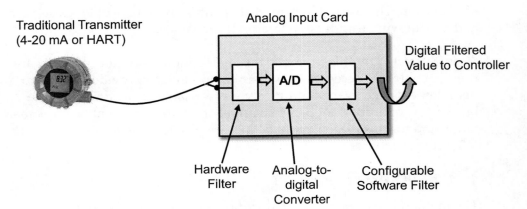

Figure 11-3. Analog Input Card Processing

The analog card may also contain a software-configurable filter after the analog to digital converter. Where this capability is provided, the software filter may be configured to provide an adjustable filter done at the card. The objective of providing software filtering in conjunction with hardware filtering at the analog input card is to prevent sample aliasing.

> **Aliasing** – Measurement distortion introduced by under-sampling an analog value.

It has been long known in communication theory that a minimum sampling rate is required to accurately process a signal. Specifically, Shannon's Theorem states that to be able to accurately represent a signal, it is necessary to sample at least two times as fast as the highest-frequency content in the signal. If a signal is sampled at a lower rate then the signal is distorted by higher frequencies being incorrectly interpreted as lower frequencies. For example, under a stroboscopic light a rotating wheel may appear to be moving backwards [2]. In analog input processes, as Shannon's Theorem predicts, aliasing will occur if the sample rate is not at least two times as fast as the highest frequency contained in the measurement signal. The distortion introduced by aliasing is illustrated in Figure 11-4 where high frequency noise is contained in the measurement signal.

In this example, the signal is changing very slowly. However, high frequency noise is contained in the measurement signal. As is illustrated, when the sampling rate is less than twice the frequency of measurement noise, then the measurement value would appear to be changing more than the actual signal and will cause inappropriate action to be taken if the measurement is used in automatic control. Thus, it is important to avoid signal aliasing.

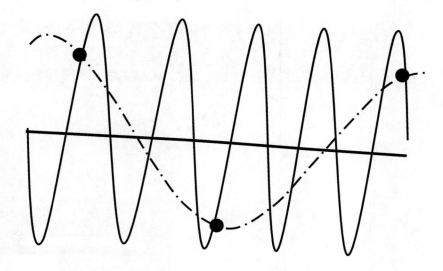

●   Value sampled by the control module

— · —   Aliased measurement as seen by the control module

————   Measurement (with noise removed)

————   Measurement with process noise

Figure 11-4. Aliasing of Measurement

The hardware filter may be designed to ensure that no aliasing occurs if control is executed at a specific rate, for example, once per 100 milliseconds. However, if control is executed at a slower rate to reduce controller loading, then added filtering may be required to avoid aliasing. A software filter that executes at the same rate as the A/D converter may be set to provide additional filtering to ensure that aliasing does not occur when control is executed at a slower rate than is supported by the hardware filter.

The software filtering on the I/O card can be adjusted to take out frequency content that is higher than one-half the frequency of control execution. Some systems allow this software filter to be accessed as part of the input channel configuration. As illustrated in Figure 11-5, by simply selecting the control execution rate, software filtering is automatically applied that guarantees aliasing does not occur.

If a measurement signal is known to be noise-free and the control execution rate is at least twice as fast as the highest frequency in the measurement, then anti-aliasing filtering is not required. For example, in many cases, a temperature measurement is relatively noise-free and temperature

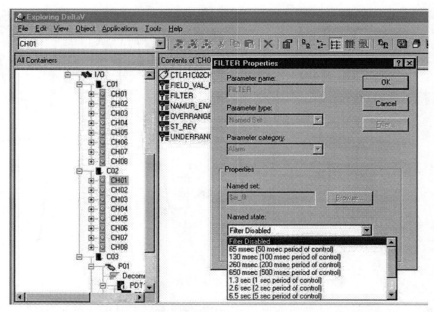

Figure 11-5. Anti-aliasing Filtering at the Analog Input Card

measurement is often slow in responding to process changes. However, other types of process inputs or outputs, such as a noisy flow measurement, may require that damping at the transmitter and/or a combination of software and hardware filtering at the analog input card be used to avoid aliasing of the measurement for use in control.

> **Damping** – Measurement filtering supported by a transmitter.

The analog input hardware and software filtering and any damping applied at the transmitter will introduce additional lag in the observed process response. From the standpoint of control, this added lag appears to be part of the process response. The lag introduced by damping and filtering is usually insignificant compared to the other process dynamics. However, on very fast liquid flow or pressure processes, filtering and damping can degrade control performance if the control execution rate is slow. Thus, anti-aliasing software filtering and added damping at the transmitter should be considered only if the measurement is noisy and not removed by the hardware filter.

If the control execution rate is set to match the filtering done by the analog input hardware filter, added software filtering at the card is not required. However, the limitations of the controller processor may require that some control be executed at a slower rate. At these slower execution rates, the hardware filter may no longer be sufficient to prevent aliasing. In general, there is a limit of how much processing is available to the controller, so control execution may be set at slower rates which will require the use of anti-aliasing software filtering at the analog input card.

11.1.3 Analog Input

The analog input block (AI block) is the basic interface to measurements used in monitoring and control applications. This block is used to provide an indication of the measurement, to provide alarming (as appropriate) upon the measurement and to provide processing of the input signal, as illustrated in Figure 11-6.

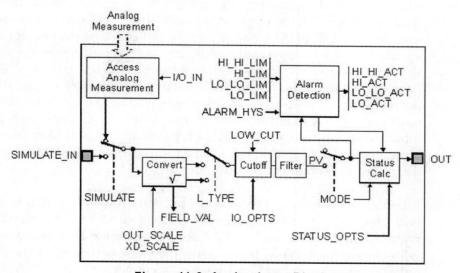

Figure 11-6. Analog Input Block

The AI block is designed to allow an input channel to be selected during the configuration of the block. This input channel corresponds to the transmitter input to an input card. Thus, through channel selection, a specific analog input may be accessed. If the analog input card is designed to work with traditional analog transmitters, then the transmitter value may be represented as a 4–20 mA signal that is accessed by the analog input block as a 0–100% value. When the input card is designed to work with a digital transmitter via a bus, such as Profibus or Foundation Fieldbus, then the digitally communicated measurement is accessed by the analog input block. When the input card interfaces to a HART transmitter, the measurement may be accessed as a 4–20 mA signal or accessed as a digital value that is communicated over the same pair of wires as carry the 4–20 mA signal.

When an analog input block is used to access a digital value, the value is typically in engineering units such as degrees Fahrenheit or gallons per minute. Thus, the analog input block must work either with measurements in percent of scale or values that are in engineering units. As illustrated above, use the L_TYPE parameter to select the type of processing done by the analog input block.

When the analog input block is working with a digital value that is already in the required engineering units, the L_TYPE parameter may be configured for Direct and the value will be used directly by the block with no unit conversion.

The 4–20 mA signal provided by a traditional or HART transmitter has some associated engineering unit range. When you are working with this type of transmitter, the engineering units' high and low values are configured in the OUT_SCALE parameter. To indicate that the percent signal should be linearly converted to this engineering unit range, the L_TYPE should be set to Indirect. When this selection is made, the 0 to 100% value provided by the input channel is linearly converted to its equivalent engineering units. The Indirect selection may also be made when there is a requirement to convert units associated with a digital engineering unit input. In such a case, the XD_SCALE parameter is used to specify the input engineering unit range and the OUT_SCALE is used to specify the desired output engineering unit range.

When an input is from a differential pressure transmitter that is installed with an orifice plate to measure flow, the square root of the signal is required. In this case, the L_TYPE parameter should be selected as Indirect with Square Root. When this is done, the value provided by the differential pressure transmitter is automatically converted to 0 to 100% of scale based on the XD_SCALE parameter. The square root of this value is calculated and then scaled to engineering units based on the OUT_SCALE parameter. As an example, a digital input of 0–50 inches H_2O might be scaled to 0–10,000 gallons/min.

As we saw in Chapter 3, a flow measurement made with an orifice plate is usable only over a certain range. Below a specific flow rate, the measurement value may not be reliable. To address this, the LOW_CUT parameter may be used to ensure a zero flow signal when the measurement is below its minimum value. This capability is commonly referred to as zero cutoff. If the measurement falls below the value specified by LOW_CUT, the value defined by the OUT_SCALE low end of range will be provided by the block.

A level transmitter may provide a noisy signal due to processing conditions such as foaming or boiling. Also, flow and pressure measurements may be noisy if the transmitter is located at a point in a process line where the liquid or gas flow is turbulent. As we have seen, the filtering provided by the analog input card may be used to remove some types of process noise to avoid aliasing. The capability to provide additional filtering is supported by the analog input block. When analog input block filter parameter, PV_FTIME, is set to zero, it has no impact on the calculated output. When the filtering parameter is set to a non-zero value, the block output response for a step change in the process input acts like a first-order process where the time constant is defined by PV_FTIME, as illustrated in Figure 11-7.

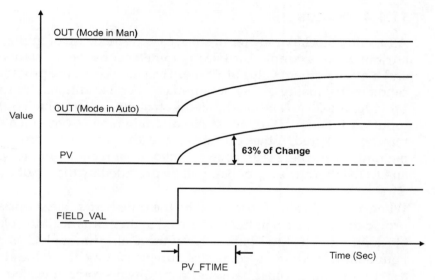

Figure 11-7. Analog Input Filtering

For example, if the analog input filtering time constant is set to five seconds, and the measurement input value changes by a step, then after five seconds the analog output will reflect 63% of its final change in value. When filtering is applied in the analog input block, the process response seen by observing the output of the analog input block will contain an added lag. From a control standpoint, the process response includes any filtering provided by the analog input block. Applying too much filtering in the analog input block can negatively impact seeing the process response. For example, applying excessive filtering may make it look like the process measurement is stable when in fact the unfiltered value may be changing significantly. Thus, excessive filtering should be avoided.

Both the analog input block and the control blocks are designed to detect and report process alarms. The alarming support by these blocks is similar in structure and configuration. Alarm detection supported by the AI block is based on the OUT parameter, that is, the filtered measurement value. An active alarm is indicated by a block parameter when the OUT value is above a high or high-high, or below a low or low-low limit. To avoid false alarms being reported when a transmitter is being calibrated, the operator may change the mode of the block to Manual. In Manual mode, the output is maintained at its last value. When the measurement input is changed during calibration, the measurement is shown in the PV (Process Variable) parameter but has no impact on the OUT parameter value. Thus, the AI block is designed to support calibration without impacting process alarming.

11.1.4 Status

A function block parameter in some cases is made up of multiple attributes. For example, the OUT parameter consists of a status attribute and a measurement value attribute. The status attribute provides an indication of the quality of the measurement. For a traditional transmitter that provides a 4–20 mA input, the status attribute is determined by the analog input card. The A/D converter has a certain range over which the current may be measured. If the current signal exceeds the A/D converter operating range (e.g., there is an open circuit or a short circuit in the wiring) then an OUT status attribute of Bad will be provided by the input card.

When a digital value is provided by the transmitter, the status attribute provided by the transmitter is reflected in the status attribute of the analog block output. The OUT parameter of the analog input block is defined by Foundation Fieldbus as a data structure consisting of a value attribute and a status attribute. These parameters are always communicated and used together.

If a transmitter is scheduled for maintenance, then the operator can "red tag" the measurement by changing the block mode to Out-of-Service. In the Out-of-Service mode, the OUT status is automatically changed to Bad – Out of Service to indicate that the measurement value should not be used in calculation or for control.

The status attribute value consists of eight bits. The most significant two bits are used to indicate the quality of the measurement: Bad, Uncertain, Good – Control, and Good – Measurement. The last two bits of the status attribute, the lower-priority bits, are used to indicate high or low limit, no limit, or constant. The four middle bits of the status attribute are used to show why the quality is good, bad, or uncertain.

Status is used in control blocks and can directly impact processing within a function block. The status option parameter is used to allow the user to determine which action should be taken under certain conditions, as indicated in Figure 11-8. Through the configuration of the status option parameter, it is possible to modify the status determination when the block is in Manual or the value is limited.

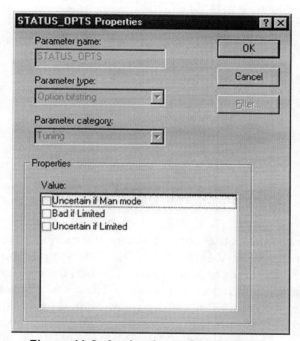

Figure 11-8. Analog Input Status Option

After status is generated by the AI block, the other blocks that use the AI output process the status and use it in some way, and so it's propagated from the AI block to the downstream blocks. In the manual loader example, with the input coming from an analog input, the status would be propagated over to the manual loader block by the connection from the AI block to the manual loader block.

11.1.5 Manual Loader Function Block

Using the manual loader function block, the plant operator may adjust the value going to the manipulated process input. The value the operator may enter as the OUT value is limited by high and low limits that are enforced by the block. As supported by the analog input block, the manual loader block allows process alarming on the filtered IN value as illustrated in Figure 11-9.

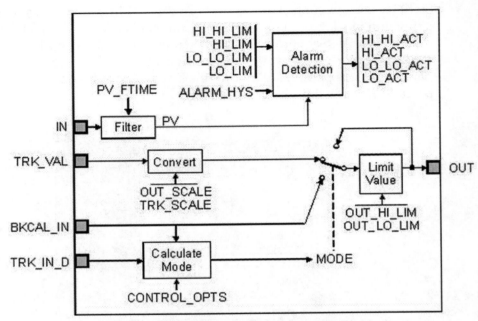

Figure 11-9. Manual Loader Function Block

Similar to many other control blocks, the manual loader function block has two inputs, TRK_VAL and TRK_IN_D to support output tracking. When the TRK_IN_D input is set to true (1) then the output of the block is set automatically to match the value of TRK_VAL. This tracking capability is often used to automate setting manipulated inputs to a defined value when the process is shut down. For example, the shutdown status of a pump may be used as the TRK_IN_D input. When the pump is not running, the manual loader block output would be positioned to the value determined by TRK_VAL. Thus, if the TRK_VAL is 30 then the OUT of the manual loader would be positioned to 30 when the TRK_IN_D has a true (1) value. Usually, a control option is used to enable the tracking feature of a block. That option must be configured before the block will take action based on the TRK_IN_D input.

11.1.6 Analog Output

The manual loader block output value set by the operator is communicated to an analog output block through a connection to the CAS_IN parameters of the analog output block. This CAS_IN parameter value is used as the setpoint of the analog output block as illustrated in Figure 11-10. This setpoint value is high and low limited.

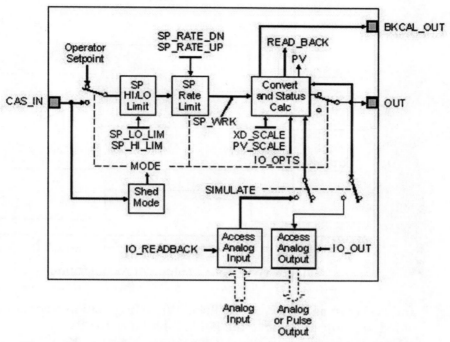

Figure 11-10. Analog Output Block Rate Limiting of Analog Output

In some processes, if the analog output is changed too quickly, then the rapid change in liquid flow rate, pressure, or temperature may damage equipment further downstream. To avoid this problem, rate limits may be imposed by the analog output block. The analog output block is the only point in a control loop where a rate limit may be imposed. Rate limiting is provided in control blocks for setpoint changes in automatic mode but this feature is disabled when the block is in Cascade mode. As the setpoint value of the analog output block is changed, then based upon the setpoint rate up or down limit, the actual value that goes out will be the ramp value as illustrated in Figure 11-11. Any time the analog output block hits a setpoint limit or limits the rate of change, the BKCAL_OUT status indicates the limit condition. The generation and propagation of limit status is used by upstream control blocks, such as the PID (proportional-integral-derivative) function block, to suspend the integral calculate when a downstream block is limited and the calculated change in integral action would exceed the output limit.

In some implementations of the AO block, an input is provided for the actuator position. If the actuator is an equipped positioner that supported digital communications, such as HART for Foundation Fieldbus, then the stem position of the actuator may be available to the analog output block

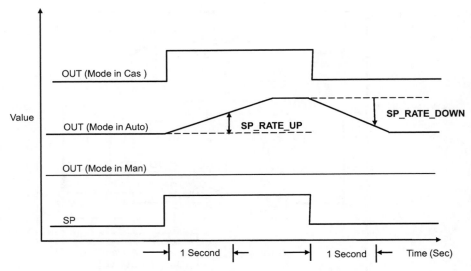

Figure 11-11. Rate Limiting of Analog Output

and is shown as the PV of the block. When no position feedback is provided, the output value to the field is shown as the PV.

The analog output block action on loss of power, that is, fail open or fail closed, is defined through the configuration of the IO_OPTS parameter as illustrated in Figure 11-12.

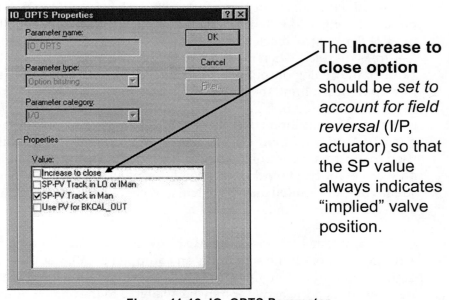

Figure 11-12. IO_OPTS Parameter

If Increase to Close is selected, 100% of output will result in an output of 4 mA, that is, the valve will fail open on loss of current or the wired connection being broken.

> **Increase/Decrease Selection** – Control system feature that accounts for valve fail-safe setup and permits the operator and control system to work in terms of implied valve position.

By correctly setting Increase to Close to match valve action on loss of current, the setpoint value will always reflect the implied valve position in the field device.

11.2 Feedback Control

Many processes have measured and unmeasured disturbance inputs that can significantly impact the value of the controlled and constraint outputs. Early in the history of the process industry, plants were compact and fairly simple. It was possible for a person to walk over the process and manually adjust valves to maintain the desired process conditions. Today, process plants are often quite complex and it is not possible for the operating staff to safely and efficiently control the process manually during normal operation.

During plant startup, when the process must be manually controlled, it is often necessary to bring in additional staff to help in bringing the process up to normal operating conditions. Once normal operating conditions are achieved, manual control isn't sufficient, so a means must be provided to allow automatic control of the processes within a plant. To permit automatic control, process control systems must provide a means for an operator to specify a desired setpoint and for the manipulated parameters to be adjusted automatically to maintain the controlled parameters at setpoint. This is done by feeding back controlled process parameters to the control system.

The most common means of providing automatic feedback control is the PID (proportional-integral-derivative) algorithm. Even though PID control has been used since before the introduction of process control systems, efforts to provide better techniques for feedback control have gained only limited acceptance. Over the years the PID algorithm has proven to be applicable to a wide variety of process applications for the implementation of feedback control and remains the heart of control systems today.

As mentioned above, feedback control is based on the principle that the process output of interest, the controlled parameter, is fed back in some manner to permit the control system to automatically adjust the manipulated process input. The simplest technique for feedback control implementation is known as single input/single output (SISO) control, as illustrated in Figure 11-13.

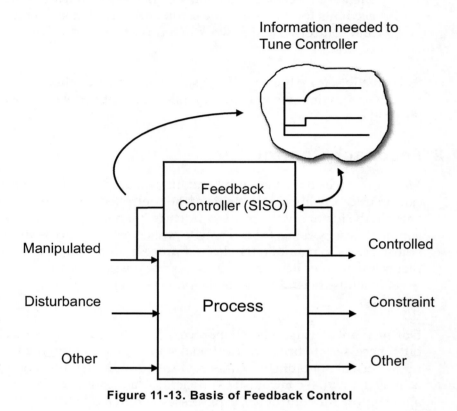

Figure 11-13. Basis of Feedback Control

In the technical community, the term SISO controller may be used to describe a controller that manages one manipulated and one controlled parameter. The PID algorithm is the most common means of implementing SISO control. As will be discussed in detail later in this chapter, certain parameters, known as tuning parameters, are used in the PID algorithm to compensate for the process gain and dynamic response associated with different processes.

> **PID tuning parameters** – Parameters of the PID algorithm that may be used to compensate for process gain and dynamic response and to tailor the control response to changes in setpoint and to process disturbance inputs.

Probably not surprisingly, it turns out that the setting of those tuning parameters is based on how the process responds to a change in the manipulated parameter. In the implementation of feedback control, the process responds based on its dynamic response and its gain, and therefore, feedback controller tuning must always be based upon the process gain and dynamic response. For a self-regulating process, the process gain, deadtime, and time constant are used directly to set the controller tuning parameters to achieve the best performance. When feedback control is applied to an integrating process, the integrating gain and process deadtime are used directly to set the controller tuning parameters. Thus, the characterization of the process response (as discussed in Chapter 9) provides information needed to implement feedback control.

11.2.1 Proportional-Only Control

The simplest application of the PID control algorithm is called proportional-only control. When only the proportional part of the PID algorithm is used in feedback control, the relationship between the controller output (OUT) and the difference between the setpoint and the controlled parameter measurement (ERROR) may be expressed as shown in Figure 11-14.

$$OUT = K_p * Error + BIAS$$

Where

OUT = Output of Controller

K$_p$ = Proportional Gain

Error = Different between the Setpoint and the controlled parameter

BIAS = bias value, also known as manual reset

P-Only

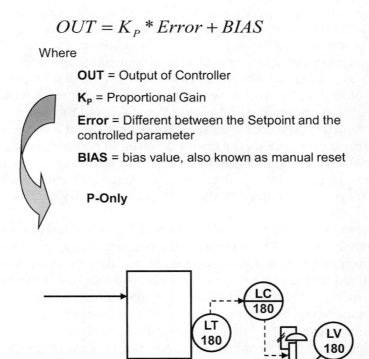

Figure 11-14. Proportional-Only Control

The output of a proportional-only controller is calculated based on a constant, called the proportional gain, multiplied by the error, plus the value defined by the BIAS parameter of the PID block. If the error changes, then the amount the output changes is determined by the proportional gain. For example, if the proportional gain was set to a value of 1 and the error was 1% of its engineering unit range, then if the BIAS parameter value were zero the output would be 1%. If the proportional gain was 0.5, then for a BIAS of zero the controller output would be 0.5%. The implementation of proportional-only control is quite simple. The controller takes immediate action on a change in error. That is, when the error changes, the output immediately changes.

A deficiency of proportional-only control becomes obvious by considering the case where the error is zero, that is, when the controlled parameter is at setpoint. For this case, the controller output would be zero if the BIAS were zero. This would seem to be a desirable condition. However, in an actual process, an output of something other than zero is required to maintain a controlled parameter at its setpoint. For example, a flow process may be designed to achieve a flow rate of 500 gallons per minute when the valve position is 70%. Using proportional-only control with BIAS set to zero, the output would be zero if the flow rate was 500 gallons per minute and the setpoint was 500 gallons per minute. However, at a controller output of zero, the valve controlling the flow would begin to close. As the valve began to close and the flow rate began to drop, an error would develop, which, when it became large enough would force the controller output to a value greater than zero and the valve would re-open. Eventually, the controller output would settle at a value where a constant error is established. The error expressed in percent of scale would be equal to the PID output divided by the proportional gain. This sustained offset between setpoint and the controlled parameter is known as "proportional droop" and is characteristic of proportional-only control.

When using proportional-only control, the amount of proportional droop may be reduced by increasing the proportional gain. However, the maximum proportional gain that may be applied for stable operation is determined by the process gain and process dynamics. If the proportional gain is increased above that value, then the control response may become unstable and become oscillatory, that is, there may be sustained oscillations in the controlled and manipulated parameters. Thus, it is not possible to eliminate proportional droop through adjustment of the proportional gain.

However, by adjusting the BIAS value of the proportional-only control, it is possible to drive the controlled parameter to setpoint. The BIAS value would be the valve position that is needed for the controlled parameter to

maintain setpoint for the current operating conditions. But, if a disturbance input changes, it will be necessary to again adjust the BIAS to drive
the controlled parameter back to setpoint. For example, consider the case
where the level of a tank is controlled using proportional-only control
where the outlet flow is manipulated to maintain the level setpoint and
the flow into the tank is a disturbance input. When the tank inlet and outlet flows are at the same value, then the level will remain constant. Thus,
by adjusting the BIAS to achieve this balance in inlet and outlet flow, the
level could be maintained at setpoint. However, if the inlet flow were to
change, then an offset would develop between the setpoint and the level
measurement. To drive the level back to setpoint, it would be necessary to
manually adjust the BIAS to compensate for the change in the disturbance
input.

For this tank level example where the PID output directly adjusts the tank
outlet flow valve, the BIAS value would be set equal to the valve position
needed to achieve normal flow rate through the tank. The proportional
gain in this example may be used to determine the outlet flow rate change
necessary to compensate for a change in level. For example, if the proportional gain is set to 1 and the BIAS is set to 50% (assuming an outlet valve
position of 50% is normally needed to maintain the tank level), then the
control output to the valve will be full open if the level reaches the upper
limit and full close if the level reaches the lower limit of the transmitter
range. If the proportional gain is set to 2 with a BIAS of 50% and a setpoint
of 50%, then the control output would be full open when the level reaches
75% and full closed when the level reaches 25%.

The proportional gain in this example determines over what range the
level is allowed to vary (float). The use of proportional-only control for
tank level control is often called floating level control. That is, the level is
allowed to vary or float within a certain range, and the amount it floats is
determined by the proportional gain. Such an approach has been taken in
the surge tank level control associated with the pulp washing area and
similar areas in a pulp and paper plant.

Proportional-only control is very simple to implement and can be used for
certain applications such as floating tank level control. In cases such as this
tank level control, the objective is to take full advantage of the surge
capacity of the tank, that is, to not immediately pass upstream process
throughput changes to the downstream process. If applied in this manner
then proportional-only control can be quite effective.

11.2.2 Proportional-Integral (PI) Control

When the control objective is to maintain the controlled parameter at setpoint under all operating conditions, then applying proportional-only

control would require that the operator constantly adjust the BIAS to compensate for changes in process disturbance inputs, defeating the purpose of an automatic control system. Because of this deficiency, proportional-only is limited to applications where the controlled parameter may vary from setpoint.

In many applications, the control objective is to automatically drive the controlled parameter to setpoint. By using the I component, the Integral term, with the Proportional term to create a PI (proportional-integral or proportional plus reset) controller, it is possible to automatically compensate for changes in disturbance inputs and maintain the controlled parameter at setpoint. The function of the integral term of the PI controller algorithm is to sum (integrate) the error over time, as shown in Figure 11-15.

$$OUT = K_P \left(Error + \frac{1}{K_I} \sum Error * \Delta t \right)$$

Where

OUT = Output of Controller

K$_P$ = Proportional Gain

Error = Different between the Setpoint and the controlled parameter

K$_I$ = Reset time, second per repeat

Δt = Period of execution (sec)

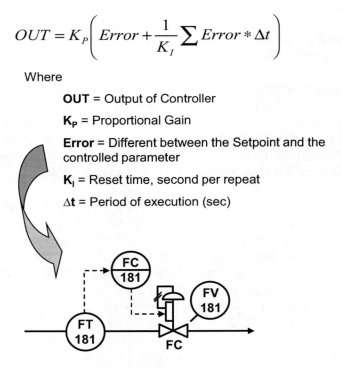

Figure 11-15. Proportional-Integral Control (PI Control)

A quick survey of the PID controllers used in the process industry would show that 90 to 95% are configured for PI control. A comparison of PI vs. the proportional-only control algorithm shows that the integral (reset) term replaces the BIAS term used in proportional-only control. In fact, the integral term may be thought of as an automatic bias that corrects for changes in load disturbances. As long as an error exists between the setpoint and controlled parameter, the summing of error (the "contribution") by the integral term continues to increase or decrease, depending upon whether the error is positive or negative, until the error is driven to zero. When the error is zero, the contribution by the integral term remains constant, that is, the integral term no longer increases or decreases. In this manner, the integral term allows the controlled parameter to be maintained at setpoint with no offset.

For example, a change in an unmeasured input disturbance will be reflected as a change in error, and as a result of the proportional contribution, there will be an immediate change in the output. This is known as the proportional kick. After that immediate response, the output will continue to change because of the integral contribution. The summing of the error by the integral term will continue until the error is driven to zero. At that point, the integral contribution will remain constant. This is how integral action automatically corrects for load disturbances, which is the fundamental advantage of using PI control for feedback control.

As mentioned above, the integral term of the PI control algorithm is also known as the *reset term*. The unit of integral gain is seconds per repeat (of proportional action). To better understand this unit, consider the case where the controlled parameter is maintained constant and the setpoint is changed to introduce an error of 1%. If the proportional gain is 1, then the change in error from 0 to 1% will cause the output to change 1% due to the proportional contribution—the proportional kick. The output will continue to change because of the integral contribution, and in the time defined by the integral gain, will duplicate the initial contribution of the proportional component. For example, if the integral gain were set to 8, then in 8 seconds the reset contribution would duplicate the initial proportional contribution, that is, a total output change of 2% would be seen for this example.

During normal plant operation, the measurement value will reflect changes in the manipulated parameter. When a setpoint change is made, a proportional kick is seen in the output and the controlled parameter begins to reflect this change and the contribution of the integral component. Eventually an output is achieved that drives the controlled parameter to setpoint. However, for explaining the units of the integral gain, it is easier to consider the case where the error is maintained constant.

Manufacturers have approached the implementation of the integral term in different ways. The most common implementation is that shown above, where the inverse of integral gain is multiplied by the sum of errors. In this case, the units of the integral gain are seconds per repeat. In other implementations, the integral gain is expressed in repeats per second and the integral gain is used to multiply the sum of errors. It is possible to convert the integral tuning from one manufacturer's implementation to another based on the units. For example, 2 seconds per repeat would be equivalent to 30 repeats per minute.

11.2.3   Proportional-Integral-Derivative (PID) Control

To provide full PID capability, a D (Derivative) component may be added to the PI algorithm. The derivative component is also known as the rate of change or "rate." This added component is based upon the rate of change of the controlled parameter and contributes to the controller output. If the rate of change in the controlled parameter is zero, then the derivative component adds nothing to the controller output. However, if the controlled parameter is quickly changing, then the derivative component will immediately add to the output in anticipation of a large change in the controlled parameter. The manner in which the derivative component contributes to the PID algorithm is illustrated in Figure 11-16.

$$OUT = K_P \left(Error + \frac{1}{K_I} \sum Error * \Delta t + K_D * Rate\ of\ Change \right)$$

Where

OUT = Output of Controller

PV = Control measurement

K$_P$ = Proportional Gain

Error = Different between the Setpoint and the controlled parameter

K$_I$ = Reset time, second per repeat

K$_D$ = Rate, seconds

Δt = Period of execution (sec)

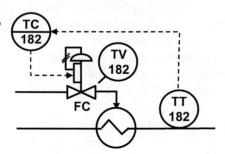

Figure 11-16. PID Control

The impact of the derivative component can be better understood by considering a few examples of changes in the controlled parameter. If the controlled parameter gradually begins to change from setpoint, then the rate of change would be small and the derivative component would add very little to the controller output. In this case, the proportional and the integral contribution would be used to bring the controlled parameter back to setpoint. However, if the controlled parameter quickly moves from the setpoint, then the rate of change is large and would significantly contribute to the controller output in anticipation of a large change in the controlled parameter. The addition of derivative action can counteract the impact of fast changing disturbances more effectively than could be achieved with just proportional plus integral action. This is the main reason that derivative action is used in control.

One of the things that limit the application of derivative action in control is that it is based on the rate of change of the controlled parameter. If there is noise in the controlled parameter, then derivative action may act on the noise and degrade, rather than improve, control performance. When the controlled parameter is a temperature measurement that is noise-free, derivative action may be used to improve control response to unmeasured disturbance inputs. However, when the controlled parameter is a noisy pressure or flow measurement, the PID control should not include derivative action. Also, some engineers avoid the use of derivative action since it adds one more PID tuning parameter that must be adjusted for best performance

11.2.4 Control Structure

Most control systems allow the user to pick the components, also known as the structure of the PID algorithm, that will be used in a control application. In some cases, this is done by providing different control blocks for P, PI, or PID. However, some systems provide a single PID block and the selection of PID components is made through a "structure" parameter. Through the structure parameter it is possible to select all combinations of the different components that may be used in PID control applications, that is, PID, PI, proportional-only, integral-only.

In the implementation of integral-only control, the proportional term is removed. Given this degree of flexibility, the natural question is, "When should I use PI versus I –only versus PID versus proportional-only?" It has been shown by Corripio and Smith that based on the characterized process response, it is possible to determine the control structure that will provide the best control performance. [3] The results of their work may be summarized in the guidelines shown in Figure 11-17.

The selection of **PID structure** should be based on the process i.e. how the controlled parameter reacts to a change in the manipulated parameter.

I-Only - When the response of the controlled parameter to a change in the manipulated parameter is instantaneous – the process is a pure gain.

PI - The process can be adequately represented as a first-order lag. *The majority of industrial process fall into this category*

PID - The process is best represented as a second-order system and the control parameter contains little noise.

P-Only - If the process is best represented as an integrator.

Figure 11-17. Guidelines in Selecting PID Structure

Most processes contain a mixture of process dynamics. In applying the guideline for selecting control structure, it is important to consider the dominant process dynamics of the control application. For example, few processes are characterized as a pure gain process. However, many processes, such as liquid flow and liquid pressure, are dominated by process gain and often exhibit negligible delay or lag in response. The response of this type of process is very close to that of a pure gain process. In most cases, the tuning of this type of process is dominated by integral action with a small proportional gain.

If the process response can be characterized as first-order, the PI structure should be selected. When the process is known to be made up of at least two dominant lags (second-order response), then PID may be more appropriate than PI control if the measurement is noise-free. For example, temperature measurement is often characterized by a second-order response. When addressing an integrating, non-self-regulating process, then selecting proportional-only or PI in which the integral term is minimized through tuning may provide the best response. These rules of thumb may be used in selecting PID structure, as well as in determining which tuning parameters should primarily be used in setting up the control system.

11.2.5 Controller Action

The process gain may be negative or positive depending on the process design. To provide the correct response a feedback control block must take into account the process gain direction—whether to increase or decrease the manipulated input to correct for a control error. For example, the inlet flow to a tank may be manipulated to maintain the tank level at setpoint. When the level increases above setpoint, then the inlet flow must decrease for the level to return to setpoint. In this case, the control response is described as reverse acting. Conversely, if tank level is maintained by adjusting the flow out of the tank, then for an increasing level, the outlet flow must increase. The control response in this case is described as direct acting. Control blocks include a parameter to select direct/reverse response as a means to accommodate this difference in control response. This selection is made internally in the block by changing the sign of the error term used in the control calculation. As part of the initial control block setup, it is important to verify that the default setting of the parameter that specifies direct/reverse action matches the application requirement. Typically the parameters used to specify direct/reverse action have a default setting of reverse since this setting is the most common. The rule shown in Figure 11-18 should be followed in configuring the direct/reverse parameter selection.

A direct/reverse selection is normally provided with the PID to compensate for the relationship of the manipulated parameter to the controlled parameter

1. Select *direct* if the manipulated parameter must be increased to correct for an increasing controlled parameter

2. Select *reverse* if the manipulated parameter must be decreased to correct for an increasing controlled parameter

   Note: The OUT parameter of the PID is normally considered to be the manipulated parameter.

Figure 11-18. Setting Controller Action

If the direct/reverse action setting is set incorrectly, then in automatic control the manipulated process input will be driven to full open or full closed. Based on the initial controller error at the time the control system was put in Automatic, the manipulated parameter will be driven in the direction opposite to that required to reduce the error. Such a mistake can be made when initially commissioning a control system, and in some cases can cause a serious upset or shutdown of a process. So, when you are initially working with a control system, it is important to always verify the configuration of the control action setting before placing control in Automatic.

11.2.6 Back Calculation

The PID control block output is commonly used to manipulate a field device such as a valve or a variable speed drive. The PID output works through a connection between the PID and the analog output block to relay changes in output to a field device. As illustrated in Figure 11-19, a second connection is required to support PID initialization and to modify the integral calculation when changes in the setpoint of the analog output block are rate or value limited.

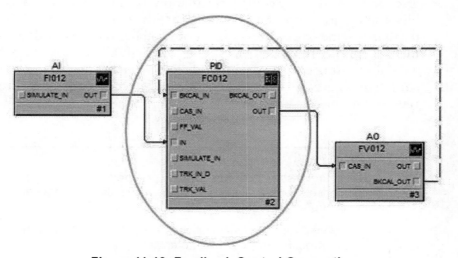

Figure 11-19. Feedback Control Connections

The mode parameter of the analog output block determines the source of its setpoint. When the block mode is set to Automatic (Auto), the plant operator determines the analog output block setpoint. In Cascade (Cas) mode, the PID output value provided through the connection between these blocks is the setpoint of the analog output block. The status attribute passed to the PID through the back calculation connection indicates when the PID output is Not Invited, that is, when the analog output block is not in Cas mode. When this status is detected, the PID output automatically tracks the analog block setpoint value provided through the back calculation connection. When the analog output block transitions from Auto to Cas mode, the PID output matches the local setpoint of the analog output block, resulting in a bumpless transfer.

The mode parameter of a control block includes a target and actual mode attribute. The desired mode of operation is specified by the operator through the target mode attribute. Based on the target mode and the status of the block inputs, the block determines the actual mode that can be supported. For example, if the target mode of the analog output block is changed from Auto to Cas, then the block's actual mode will not transition to Cas until a handshake is completed between the PID and the analog output block. This handshake consists of the following steps:

1.   When the analog output block target mode is changed to Cas, the status in the back calculation connection changes from Not Invited to Initialization Request.

2.   On detection of Initialization Request in the back calculation input status, the PID sets its output to match the analog output block setpoint value provided through the back calculation connection. Once initialization is complete, the output status is set to Initialization Acknowledge.

3.   On detection of Initialization Acknowledge in the PID connection, the analog output block changes the actual mode attribute to Cas. It then starts to use the PID output provided through the connection input as its setpoint.

The use of a handshake between blocks ensures that the PID output matches the analog output block setpoint when the actual mode of the analog output block transitions from Auto to Cas mode. If the blocks are configured to execute at a fast rate then the handshake between blocks may occur so fast that it may not be apparent. Nonetheless, it is a very important feature of the function blocks used in control applications. Its correct operation in providing bumpless transfer depends on the back calculation connection being properly configured.

The information provided in the back calculation connection continues to be used during normal operation. For example, if the rate of change or value of a setpoint exceeds limits specified in the analog output block, then a limited value will be used to adjust the field device(s). The high and low limits status and limited setpoint value passed through the back calculation connection are used to alert the PID that a change request was not fully acted on by the downstream block. For example, if the setpoint high limit is set to 60% in the analog output block, then a PID output requesting a setpoint greater than 60% will result in a value of 60% being used in the analog output block. When this limit is hit, the status in the back calculation connection is set to indicate a High Limit condition.

The PID is designed to compensate for limiting in a downstream block. If the error is such that the reset is going to drive the PID output further into the limit condition, then the last value of the reset contribution may be maintained or calculated, based on the limited value, to avoid reset windup.

Reset windup – Continued PID accumulation of reset contribution when further changes in the PID output will have no impact on the process.

Unless the PID block takes anti-reset windup action, the integral contribution can accumulate to a large value. If conditions then change and it becomes necessary for the controller output to move in the opposite direction, it will be necessary to "unwind" the accumulated reset before the controller output can move in the opposite direction. Thus, reset windup can introduce a significant delay in responding to process changes. For this reason, the PID is designed to automatically provide anti-reset windup. Normal integral action is restored when changes in the PID integral contribution to the output will move the output away from a limit condition.

11.3 PID Block Implementation

PID block implementation may vary depending on the manufacturer. However, the common functions performed by the PID block may be illustrated as shown in Figure 11-20.

In addition to the input for the controlled parameter, other inputs are normally provided that are needed to support the use of the PID block in multi-loop control (which will be discussed in Chapter 13). Also, PID inputs to support output tracking are commonly provided to allow the PID output to be automatically positioned to a fixed value under certain conditions, such as when the process is shut down.

The heart of the PID block is the PID algorithm acting on the controlled parameter. The settings of the GAIN, RESET, and RATE parameters determine the response of the PID in automatic control. As is illustrated in this Figure 11-20, the manufacturer of the PID block may include added optional parameters that can impact PID block operation. For example, the option may be provided to automatically decrease the proportional gain when the controlled parameter is close to setpoint. These types of PID options may be useful under some situations, but for most applications they are not required and thus are not addressed in this chapter. The manufacturer's documentation should be referenced to learn about the optional manufacturer-specific options included in the PID block.

An overview of the implementation of a PID block is normally provided by the manufacturer. The actual implementation of a PID block is often quite complex. The design must support both automatic and manual control, added inputs used for multi-loop control, and functionality associated with control options that the manufacturer may have included in the

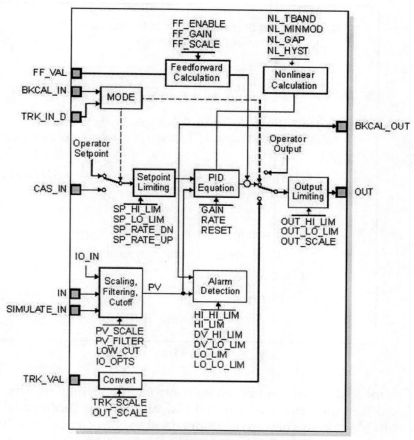

Figure 11-20. PID Block Implementation

PID block. The basic concepts of PID control are fairly simple, but there are many details that must be addressed in the block implementation.

Different approaches may be taken in the implementation of the PID algorithm. In industrial process control systems, the most popular approaches are known as the standard form and the series form. A manufacturer may show the PID algorithm implementation of the standard and series from using the Laplace transform (frequency domain) representation as illustrated in Figure 11-21.

The reason for this is that most often the block design is more precisely described using the frequency domain representation. The implementation of the PID block may be quite difficult to concisely document since it must include details such as output limiting. Also, the implementation may include details concerning control options that are proprietary to the control system manufacturer. Independent of the manufacturer's implementation of the PID block, the basic concepts of feedback control are the

Conventional Standard PID with feedforward
– Laplace (s domain) representation.

$$OUT(s) = GAIN * \left(1 + \frac{1}{T_r s} + \frac{T_d s}{(\alpha T_d s + 1)} \right) * E(s) + F(s)$$

Series PID with derivative filter applied only to derivative action, with feedforward – Laplace (s domain) representation.

$$OUT(s) = GAIN * \left(1 + \frac{T_d s}{(\alpha T_d s + 1)} \right) * \left(\frac{T_r s + 1}{T_r} \right) * E(s) + F(s)$$

where:

$GAIN = \mathrm{Pr}\,oportional\ gain$

$T_r = \mathrm{Re}\,set\ time,\ \sec onds\ per\ repeat$

$T_d = Rate,\ \sec onds$

$E(s) = Error$

$F(s) = Feedforward\ contribution$

Figure 11-21. PID Form

same and set the foundation for control implementation. The many details of PID implementation must be addressed by the control system manufacturer but do not need to be understood to correctly use the PID block in control applications.

11.3.1 PID Form and Structure

As mentioned above, the two most common implementations of the PID block in process control systems are the standard and series form of the PID algorithm. If derivative action is not required, that is, the RATE parameter is set to zero, then the standard and series forms of the PID are identical in the way the calculation is done. If rate action is used in a control application, then there's a subtle difference between the ways the derivative component is added in the standard and series implementation. In the series form, the proportional and integral contributions are multiplied by the derivative contribution. In the standard form, the derivative contribution is added to the proportional and integral contributions.

There are pros and cons to each approach. One advantage of the series implementation is that adjustment of the derivative contribution will have no impact on the proportional and integral contributions. Since both implementations are common in the process industry, some manufacturers of process control equipment allow the end user to select the PID form that will be used in the PID block. For example, the manufacturer may provide a PID control block option that may be used to select the PID form, as illustrated in Figure 11-22.

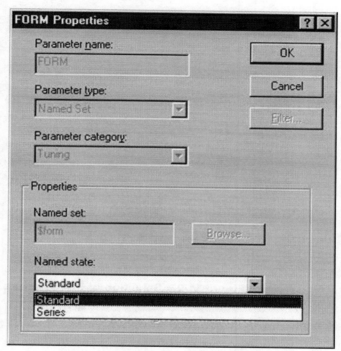

Figure 11-22. Selection of PID Form

As discussed in 11.2.4, depending on the application, the feedback control requirement may be best addressed using proportional-only control, integral-only control, proportional with reset control, or with proportional-integral-derivative control. Some manufacturers provide multiple blocks to allow the user to select the desired type of feedback control. Alternatively, one block may be provided with a parameter that may be configured to select the structure of the PID.

This structure option may also allow the user to select whether the derivative contribution is based on the error or on the controlled parameter value. For many applications, it is desirable for derivative action to be based on the control measurement rather than on the error. When the derivative is based on the control measurement, it is possible to avoid a spike in control output when the setpoint is changed. An example of how the PID block structure may be selected is illustrated in Figure 11-23.

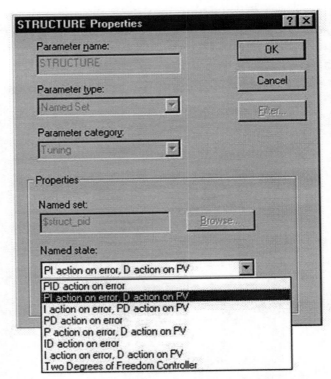

Figure 11-23. Selection of PID structure

As with the derivative contribution, when the option is to base proportional action on the controlled parameter rather than on the error, a sudden change in the output due to a setpoint change may be avoided. In this case, the initial response to a setpoint change is provided by the reset contribution to the calculated PID output. When the setpoint is changed, the output is changed in a controlled manner to bring the controlled parameter to setpoint. As a result, the response to setpoint changes will be slower than when the structure is selected for proportional action on error.

This flexibility in selecting the control structure complicates the manufacturer's implementation of the PID. However, such flexibility has the benefit of allowing the user to use one PID block for all feedback applications. Any changes in the control structure can thus be easily implemented through the structure selection.

11.3.2 Mode

The mode parameter of the PID block determines the source of the block setpoint and the source of its output. This parameter is a critical part of the operator's interface to the PID block and consists of four attributes:

Target – The mode of operation requested by the operator

Actual – The mode of operation that may be achieved based on the target mode and the status of the PID block inputs

Permitted – The mode(s) available to the operator for a given application

Normal – The mode the PID is expected to be in for normal operation

When the block is set to Automatic (Auto) mode, the setpoint of the block is determined by the operator while the block output is automatically regulated by the block to maintain the controlled parameter at setpoint. In Cascade (Cas) mode, the setpoint is provided by another function block through the CAS_IN connection. By changing the block mode from Auto to Cas, the source of setpoint is changed to be the CAS_IN value. The value written to the RCAS_IN by an external application will be used as the setpoint when the mode is set to Remote Cascade, RCas. For example, in a batch application, the batch control logic may change the setpoint of a PID block by writing the setpoint value to the RCAS_IN parameter and then changing the block mode to RCas. The selection process supported by the mode parameter is illustrated in Figure 11-24.

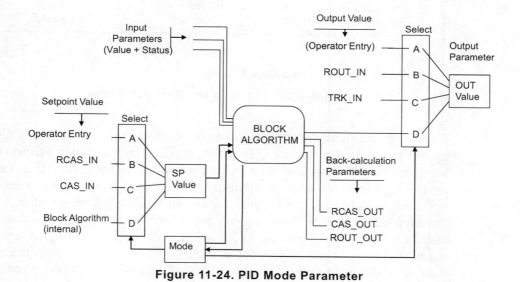

Figure 11-24. PID Mode Parameter

A batch control application may constantly write to the RCAS_IN parameter but this value will only be used as the setpoint of the PID when the mode parameter is changed to RCas. The operator may take back control of the setpoint at any time by changing the block mode to Auto. Thus, the mode parameter is used by the operator to determine when it is appropriate for a batch control application to adjust the PID setpoint.

When the PID block mode is changed to Manual (Man), then the operator may set the output of the PID block rather than using the calculated output. In Man mode, the PID output is maintained at the value specified by the operator. When there is a need for an external application to take control of the PID output, then the operator may change the mode of the block to Rout. In the Rout mode, the value written to the ROUT_IN parameter will be used as the PID block output.

Through the configuration of Permitted attributes of the mode parameter, it is possible to specify which modes may be selected by an operator. The operator modes for the PID and the impact of the mode selection on the source of setpoint and output are summarized in Figure 11-25.

Mode	Source of SP	Source of OUT
Out-of-Service(O/S)	Operator	Operator
Manual (Man)_	Operator	Operator
Automatic (Auto)	Operator	Block
Cascade (Cas)	CAS_IN	Block
Remote Cascade(Rcas)	RCAS_IN	Block
Remote Output (Rout)	Operator	RCAS_OUT

Control and output blocks

Figure 11-25. Supported Operator Modes

The Actual mode attribute may take on a value other than that specified by the Target mode attribute. This occurs when an internal condition or input status indicates that the mode of operation requested by the operator through the Target mode attribute can not be achieved. Under normal operating conditions, the Target and Actual mode attributes match. The Actual mode attribute changing to Local Override (LO) indicates that the output tracking or an internal application, such as auto-tuning, is setting the block output. Similarly, if the path to the process is lost, then the Actual mode attribute will change to Initialization Manual (IMan) as summarized in Figure 11-26.

Actual Mode	What it means
Local Override (LO)	Track or Auto-tuning is active and in control of the output value
Initialization Manual (IMAN)	The forward path to a physical output is broken and the output is tracking the downstream block

Figure 11-26. Other Actual Modes

When the Actual mode of a PID is shown as IMan, this indicates that the path to the process is not complete. The path to the process can be broken if a downstream block is taken out of Cascade mode. For example, in a single-loop application, the path of the PID output to the process will be broken by changing the mode of the analog output block from Cascade to Automatic. The path to the process may also be broken by a physical condition such as the failure of an actuator that is manipulated by the PID. When the Actual mode of the PID is IMan, the PID cannot adjust the manipulated process input. Under this condition, the PID output follows the value provided by the downstream block through the back calculation input.

11.4 Pulsed Outputs

The process control examples presented thus far have shown regulating valve position as the process manipulated input. The AO block has been the means of interfacing to the regulating valve. However, field devices such as damper drives, on-off valves, and electric heaters that are used in the process industry may require a pulsed output interface to manipulate the process input. This section addresses some of the most common types of pulsed output interfaces. Examples are used to illustrate how these may be used in single loop control applications.

11.4.1 Duty Cycle Control

One means of providing heat input to a process is through electric heater bands. As with an electric blanket, heater bands contain resistance coils that heat up due to an electric current. For example, in the plastics industry an extruder is used to process plastic pellets into various products. Plastic pellets are fed into the barrel of the extruder, where they are heated using electric heater bands. As a close-fitting screw turns inside the barrel, the pellets move through the extruder and melt, and the liquid plastic is forced out through a shaped opening known as a die to form a final product, such as a plastic sheet. Because melting the plastic requires energy, energy must be supplied to the extruder. To maintain the temperature of the liquid plastic at a desired setpoint value, it is necessary to adjust the energy input to the extruder by regulating the current flow through the heater bands.

In the case of the extruder, no valves are involved in regulating the process input, which is electrical energy. Instead, the energy input to the extruder is manipulated by interrupting the current flow to the heater bands. A 220-volt AC circuit is often used to power the heater bands. When voltage is applied, a current flow is established based on the resistance of the wires that make up a heater band. The heat energy that is provided by the heater band over a specific period of time is determined by the heater band resistance, the voltage of the circuit, and how long the voltage is applied. This regulation of energy input by turning the voltage on and off is known as duty cycle control.

> **Duty cycle control** – Regulation of a process input by varying the percentage of the time the process input is turned on over a specific period of time known as the duty cycle.

In the case of the extruder heater band, energy input is regulated by applying voltage to the heater band for a fraction of the duty cycle, and then removing the voltage for the remainder of the control period. If the amount of ON time is zero, then no energy will flow to the heater band, and therefore, there will be no impact on the barrel temperature. If the fraction of ON time is increased to one second out of every ten seconds, then 10% of the maximum energy input will be applied to the extruder barrel and the energy input will increase. If the ON time is changed to two seconds out of ten, then 20% of the maximum energy input will go to the barrel. When power is applied continuously to the heater band, the maximum energy input to the process will be applied.

As mentioned, heat energy is lost during extruder operation and must be replaced by the heater bands. Thus, the barrel temperature can be maintained at setpoint by adjusting the ON time over the duty cycle period. In the application of duty cycle control, the duty cycle time should be set at a value that is small compared to the process response time. For example, because of the extruder barrel mass, the temperature response to a change in energy input may be characterized by a time constant of one or two minutes. For a duty cycle of ten seconds, the changes in temperature resulting from power being turned on and off during the duty cycle would be filtered out, resulting in a constant barrel temperature.

To provide duty cycle control, process control systems may include an interface to regulate a discrete output. For example, the discrete output card may be designed to allow a channel to be configured for duty cycle control. The AO block may be designed to work with a discrete channel. When this flexibility is provided, the AO block regulates the ON time associated with the discrete output channel. For example, if the AO block setpoint is changed to zero, then the discrete output will remain zero; it never turns on. However, if the AO block setpoint is changed to 50% then the discrete output will be turned on for 50% of the duty cycle time and then turned off for the remaining portion. Thus, the setup for PID control when the process input is regulated using duty cycle control is the same as that used when the process input is regulated by a valve. The only difference is in the setup of the AO and the discrete output card. When the discrete channel is set up for duty cycle control, the timing of the discrete ON and OFF time over the duty cycle is done by the discrete card as illustrated in Figure 11-27.

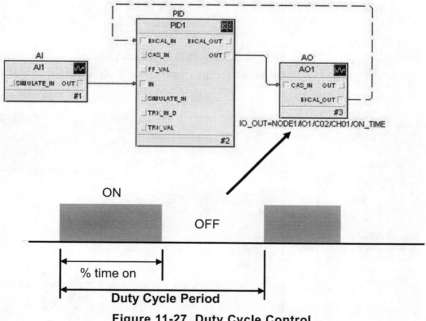

Figure 11-27. Duty Cycle Control

The timing resolution of the ON and OFF time must be precise. For example, the timing resolution may be one-half of one cycle of 60-Hertz power, which is 1/120 of a second resolution. Such resolution of power allows very precise temperature control. The setup of a discrete channel for duty cycle control is shown in Figure 11-28.

When using the AO block to reference a discrete output that is configured for duty cycle control, the user may reference the discrete channel attribute ON_TIME as illustrated in Figure 11-29.

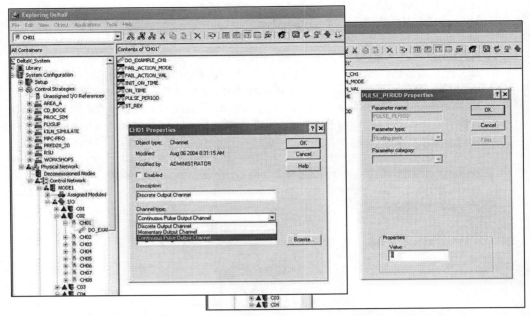

Figure 11-28. Discrete Output Setup for Duty Cycle Control

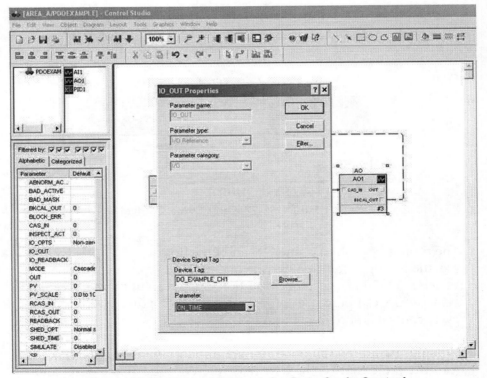

Figure 11-29. Using AO Block for Duty Cycle Control

Duty cycle control also may be applied in the chemical industry to regulate on-off valves that are used to meter small amounts of material into a tank, and in the life sciences industry, duty cycle control is used to regulate the on-off state of peristaltic pumps that are used to provide feedstock to a bioreactor.

11.4.2 Increase-Decrease Control

Motorized actuators may be used for applications that require larger valves, such as those used for the regulation of wastewater flow. Also, a motorized actuator may be preferred in some applications, such as the basis weight control of a paper machine, that require precise positioning of the valve. In such an application, these actuators typically have some gearing mechanism and the motor is designed to run in a forward or reverse direction. This allows a valve to be opened or closed by running the motor in a forward or reverse direction. The length of time the motor remains on, and its direction, determines how much a valve is closed or opened. In these types of applications, it is possible to use a discrete output that supports duty cycle control to tell the motor when to turn on and off. However, since the motor can run in a forward or reverse direction, it is necessary to use two discrete outputs to interface to a motorized actuator. This type of regulation is known as increase-decrease control.

> **Increase-decrease control** – Regulation of a motorized valve actuator through the use of discrete contacts to run the motor in a forward or reverse direction.

The implementation of increase-decrease control is simplified when a measurement of the actuator position, such as valve stem position, is available for use in the discrete output interface. In this case, the regulation of the discrete outputs for forward and reverse operation can be based on the error between the position setpoint and the measured position. For example, if the error is positive then the motor may be run in the reverse direction for a period of time based on the magnitude of the error. Similarly, if the error is negative, then the motor may be run in a forward direction for a period of time based on the magnitude of the error. The calculation of the error and the coordination of the two discrete outputs can be implemented using a calculation and splitter block as illustrated in Figure 11-30.

In this example, the AO blocks would be configured to work with discrete output channels that are set up for duty cycle control. Since there is some minimum on-time that the motor will respond to, it is necessary in the implementation to output changes that can be acted on and to integrate any changes that cannot be acted on by the motor.

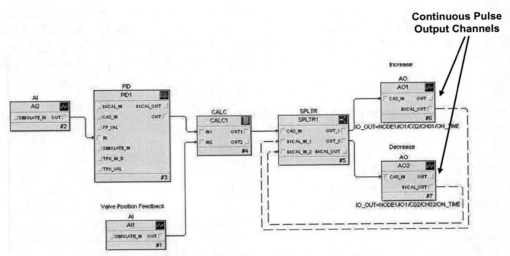

Figure 11-30. Interfacing to Increase-Decrease Actuators

11.5 Process Action

The PID block is normally designed to support added inputs for tracking to enforce process action.

> **Process action** – Mechanism for connecting or coordinating the functions of different PID components.

If a condition is indicated by a discrete input that requires process action, then this input is provided as a tracking input to the PID. When the tracking input changes to an ON state, the PID will position its output to a fixed position and maintain that value as long as the tracking input remains active. The tracking position may be specified as part of the PID block configuration, or in some systems may be defined by an input to the PID block.

The discrete input that is used to trigger tracking may indicate the run state of a motor. For example, if a feed pump is shut down then it may be appropriate to close the valve that is used to regulate the pump flow. By using the status of the pump motor as the PID tracking input, the valve can be automatically closed when the pump is not running.

When setting up tracking on the PID, it may be necessary to enable this feature of the PID. Also, as illustrated in Figure 11-31, there may be options that may be used to select whether the tracking function is valid when the PID mode is Manual.

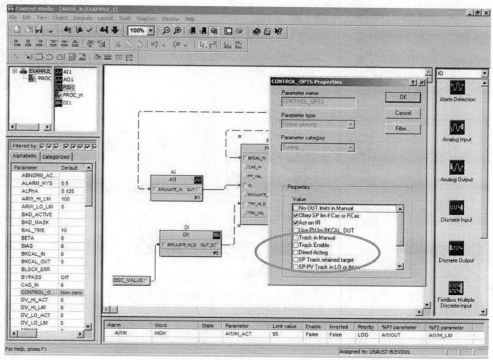

Figure 11-31. Configuring Tracking to Enforce Process Actions

In setting tracking options, it is necessary to consider the circumstances under which the operator may need to override tracking by placing the PID in Manual mode. If the option is selected to allow tracking in Manual, then when a condition is active that requires process action, the operator will not have the ability to manually position the PID output. If the operator is trusted to make the right decision, then the option to track in Manual mode should not be selected. If the interlock that is being enforced using tracking is to protect equipment, that is, the process action must be observed to avoid a condition that would damage equipment, then tracking in Manual may be selected.

11.6 Workshop – Feedback Control

This feedback control workshop is designed to allow you to explore and become more familiar with many of the concepts that have been introduced in this chapter. The control of a temperature process using the PID block is the example used in this workshop. The response of the process to a step input change can be observed by placing the PID block in Manual and changing its output through the OUT parameter. The AO block mode may be changed from Cascade to Auto to see its impact on PID mode and operation. You are encouraged to use this exercise as a means of getting hands-on experience in using the PID block for feedback control.

The temperature process that is simulated for this feedback control workshop is illustrated in Figure 11-32.

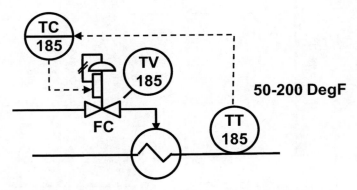

Figure 11-32. Process for Feedback Control Workshop

See the Appendix for directions on using the web interface.

Workshop Directions:

Step 1: Using the web interface, open the Feedback Control workshop. Select the workspace tab shown in Figure 11-33, set the mode of the PID to Manual using the MODE parameter. Change the PID output using the OUT parameter and observe the process response.

Step 2: Set the PID mode to Auto and change the setpoint using the SP parameter. Observe the process response.

Step 3: Introduce an unmeasured process disturbance by changing the DISTURBANCE input. Observe the impact on the process and the action taken by the PID block to return the temperature to its setpoint.

Step 4: Reduce the PID GAIN by a fact of 2 and then repeat steps 2 and 3. Observe the impact of the gain change on the response to setpoint and disturbance changes.

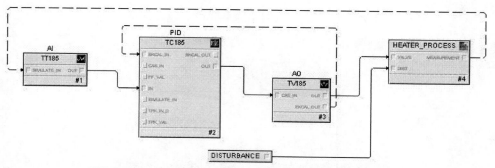

Figure 11-33. Feedback Control Workshop Module

References

1.   Fieldbus Foundation, http://www.fieldbus.org/

2.   Kostic, M. "Sampling and Aliasing: An Interactive and On-Line Virtual Experiment," ASEE 2003 Annual Conference and Exposition.

3.   Corripio, A. B. and Smith, C. L. "Mode Selection and Tuning of Analog Controllers." Paper presented at the 3rd ISA Biannual Instrument Maintenance Clinic, Baton Rouge, Louisiana, 1970.

12

Tuning and Loop Performance

During the commissioning of feedback control based on the PID algorithm, the tuning parameters associated with the PID (that is, the proportional, integral and derivative gain) must be set to specific values to achieve the best controller response to setpoint and disturbance input changes. The PID tuning parameters are normally specified as positive, floating-point values including zero and can take on many combinations of values. The procedure used to set the PID tuning parameters is commonly referred to as loop tuning. In this chapter, insight is provided into techniques that may be used to set the PID tuning parameters for best control performance. Also, some of the things that may impact loop tuning are addressed.

12.1 Initial Loop Tuning

During the design for a new process area or plant, it may be necessary to define the initial loop tuning that should be used when configuring the control system. At this state, very little may be known about the process gain and dynamics associated with a control loop. If the plant has not yet been built then there is no opportunity to characterize the process response by observing the response to step changes in process input. All that may be known is the type of measurement associated with the control loop, that is, whether pressure, level, temperature, flow, or some other process variable is to be controlled. In this case, the typical tuning parameter values shown in Figure 12-1 may be used based on the measurement type.

In most cases, these default settings will be close enough to support the initial startup of a new process area or plant. In addition, the settings based on this table will be conservative and will provide a stable response

	Gain	Reset	Rate
Flow	0.3	5	--
Temperature	1.3	300	60
Level	2	600	--
Gas Pressure	3	600	--

Figure 12-1. Initial PID Tuning

to setpoint and load disturbance changes during automatic control. In fact, it may not be necessary to change the initial default tuning during plant startup if stable operation is achieved and the time for the control to compensate for a change in setpoint or disturbance input is adequate to meet the plant operating requirements. However, to minimize process variations from setpoint and the respond time for setpoint and disturbance input changes while providing stable operation over a variety of operating conditions, during or after startup, it is necessary to set the PID tuning based on the observed process gain and dynamics for each control loop. Unfortunately, the instrument engineer involved in plant commissioning or plant operations may not have the opportunity to conduct loop tuning and as a result, the plant operation does not achieve maximum efficiency.

In an operating plant, various approaches may be taken by plant personnel to modify control loop tuning to improve control performance. In some cases, there may not be a good understanding of how to set control tuning based on the process response. In such cases, the person involved may have a "little black book" or some other means of recording tuning that has previously worked on various types of loops. For a flow loop, a favorite tuning may be recorded and will be applied when attempting to improve the performance of another flow loop. Because this approach may not take into account the differences between the processes, using such an approach to tuning may not result in improved control performance as measured by the time to respond to setpoint and load disturbances and variations from setpoint.

At that point, plant personnel may revert to a seat-of-the-pants approach in which the loop tuning is changed by trial and error until some combination of tuning parameter values is found that seems to improve control performance. Such an approach is very time consuming and the tuning results often differs from that required to achieve best control performance.

As stated above, a much more direct and reliable way of establishing loop tuning is to manually or automatically set the PID block tuning based on the identified process gain and dynamics as described in the following sections.

12.2 Manual Tuning

With older control systems, the analog single loop controllers and programmable logic controllers (PLCs) used may not include tools that allow the PID tuning to be automatically established based on the observed process response to input changes. When working with such systems, it is necessary to introduce a step change in a manipulated input with the PID controller or PID block in Manual and then observe the response of the controlled parameter. With modern control systems, the manual establishment of PID block tuning can also be useful when initially commissioning process areas since this allows the tuning to be quickly established and control to be placed in Automatic during plant startup.

When commissioning PID control associated with a self-regulating process, the procedure outlined in Figure 12-2 may be quickly applied to both old and new control systems to determine the tuning for PI control.

Tuning of a PI controller applied to a self-regulating process can be quickly establish as follows:

1. Place the controlled and manipulated parameters on trend.
2. Place the controller in manual and allow the process to reach steady state.
3. Impose a step change in OUT and observe the response.
4. Set the RESET to match the sum of the process deadtime plus the time constant.
5. Place the loop on automatic control using conservative GAIN.
6. Make small changes in Setpoint and observe the response. Adjust only the GAIN to achieve the desired response.

Figure 12-2. Manual Tuning Technique

The size of the step change introduced in the process input depends on the process gain. In general, the step size should be just large enough to easily distinguish the resulting change in the controlled process output. Based on the process response and the size of the step change as previously discussed in Chapter 9, the process gain, deadtime, and time constant may be determined. The reset should be set to a value equal to the sum of the process deadtime and the time constant.

RESET Units	RESET Value
Seconds/Repeat	Time Constant + Deadtime (sec)
Repeats/Minute	60 /(Time Constant + Deadtime (sec))

For example, if the total response time (time constant + deadtime) is 15 seconds and the RESET units are seconds/repeat then the Reset would be set to 15. With the Reset being set, for PI control it is only necessary to determine the proportional gain setting of the PID. A very conservative gain, for example, 0.2 should initially be selected and the PID mode then changed to automatic control. Once the PID is placed in automatic control, the setpoint may be changed. If the response to the setpoint change is slower than desired, then the proportional gain should be increased by a small amount (e.g., 20%) and the setpoint changed again. If the response is still too slow then the proportional gain should be increased again. This process of changing the proportional gain and observing the response to a setpoint change should continue until the desired response is achieved.

The response of the controlled parameter to a setpoint or load disturbance change may be classified in the following manner:

Over-damped response – Controlled parameter is gradually restored without overshooting the setpoint value. In most cases, an over-damped response provides best response for varying operating conditions that impact process gain and response time.

Critically damped response – Controlled parameter is restored in minimum time without overshooting the setpoint value. While minimizing response time, unstable control may be observed with changing operating conditions that impact process gain or response time.

Underdamped response – Controlled parameter overshoots the setpoint value but eventually settles at the setpoint value. May minimize time to get back to setpoint but at the expense of stability and overshoot to setpoint and load disturbances.

Thus, just by increasing the PID proportional gain, it is possible to go from an overdamped response to a critically damped response to an under-damped response. The reset value is unchanged since it must be set based on the process time constant and deadtime. Once the reset is set, then the desired response may be achieved just by adjusting the PID proportional gain.

If the PID proportional gain is increased beyond the value required for an underdamped response, then on a step change in setpoint the PID output and controlled parameter may break into a sustained oscillation. At this point, the response to a change in setpoint or disturbance input is classi-fied as unstable.

> **Unstable response** – Controlled and manipulated parame-ters exhibit an oscillatory response that builds in amplitude until the manipulated parameter reaches its output limits.

An unstable control response may cause significant operational problems in a plant environment. To avoid an unstable control response, when tun-ing the PID it is important to change the gain in small steps. It is always best to start with a conservative gain and then test the control response in Automatic after each increase in gain. Using this technique, it is possible to tune control loops associated with fast responding processes in a matter of minutes.

When you are tuning control loops, natural questions might include, "What is the best tuning for a particular loop? Should loop control perfor-mance be measured based on the observed process response for setpoint change, or to a change in a load disturbance? What factors should be con-sidered when judging loop control performance?"

The answers depend on what the objective is for the loop. If the setpoint is constantly being changed then the response of the loop to setpoint changes would be a key factor in evaluating its performance. However, if the loop setpoint is always maintained at a fixed value, then the response to disturbance inputs would be most important. If the response to both setpoint and disturbance inputs is important, then tuning must be adjusted to achieve a balance in control response.

In many processes, it is desirable to have an overdamped response. To achieve that response, the proportional gain used in the PID is typically much less than that required for an underdamped or unstable response.

When selecting the PID gain, it is important to consider that the process gain may change with the operating conditions. For example, the installed characteristic of a valve may provide a non-linear response to valve position in the valve operating range. The PID gain should be based on the operating conditions that provide for maximum process gain. If the process gain changes, it has the same effect on the control loop response as changes made in the PID gain.

A mistake that is often made in tuning control loops is to examine loop control performance at one operating point. For example, when a valve is operating at approximately 50% open, tuning may be adjusted to achieve an underdamped response. When the setpoint is changed to where the valve is operating at 30% open an entirely different response may be observed because the process gain has changed. As has been emphasized, it is wise to select a conservative PID gain setting which will provide stable control even if the process gain changes with the setpoint. As a consequence, the "best" PID gain often gives a slower response to changes in setpoint and load disturbances. This conservative tuning will help ensure stable operation even if the process gain changes with increases or decreases in the operating setpoint.

The need to consider the full process operating range when tuning a loop can be illustrated by an example. The portion of a control system associated with the temperature of the steam leaving a boiler is to be commissioned. The control performance criterion is for the loop to respond as fast as possible to setpoint changes. However, in this example, let's assume that testing and commissioning of the loop was only done at one setpoint and when the setpoint is increased, the response is underdamped. Further increases in the setpoint may result in the loop going unstable. In this example, as the setpoint is increased, the valve position needed to maintain the temperature changes from 50% to approximately 40%. Further increases in the temperature setpoint require a valve position of only 30% to maintain temperature. When the steady-state temperature vs. the valve position is plotted, it reveals that the process gain at a 50% valve opening is significantly less than the gain at a 40 or 30% valve opening. These changes in the process gain directly impact the closed loop response since the product of the proportional gain and process gain determines the closed loop gain. With each increase in the operating setpoint, the valve moved into a region of higher process gain. At this higher gain, an underdamped response was observed.

To ensure stable operation over the entire operating range of a control loop, the normal practice is to establish the PID tuning when the process is operating in the region with the highest process gain. If the loop operating point changes to a low-gain region, the worst that happens is the response to a setpoint change is sluggish, but the response is stable. If a slower response does not meet process requirements, then it may be necessary to use a characterizer to provide linear installed characteristics as discussed in Chapter 9.

When the process gain changes with operating conditions, the best control performance over the entire operating range would be achieved if the proportional gain were automatically changed based on the process gain at each point of operation. Some control systems provide a proportional gain scheduling capability that may be used to change the PID gain based on valve position or some other measured or calculated parameter that is related to the process gain. The setup of this gain scheduling capability is time consuming since the process gain must be established over the entire control operating range. So, gain scheduling may not be used or may be limited in use to a few processes that are highly non-linear and cannot be adequately controlled by simply establishing the PID gain in the high process gain region.

12.3 Automatically Establishing Tuning

Many of the newer control systems used in the process industry include auto-tuning, the capability to automatically establish PID block tuning. This capability is based on automatically analyzing the on-line controlled parameter response to changes in the manipulated process input. Control system manufacturers have adopted different technologies to implement auto-tuning. Since these tools are designed to work in a noisy process measurement environment, the tuning results may be better than those achieved through manual tuning techniques. As an example of an application of auto-tuning, the auto-tuning capability tool by one manufacturer will be examined in this chapter. Other tools may differ in the requirements to test the process and the methods utilized to identify the process gain and dynamics.

Auto-tuning is designed to automatically determine the proportional, integral, and derivative gains and thus can potentially save time in commissioning a control loop. Since this capability will be used in many process applications, it must be designed to work with a variety of processes: self-regulating, integrating, fast or slow response. Protection is typically provided to compensate for noise in the measurements and the influence of unmeasured load disturbances. Also, to be useful in initial plant startup, it is important that the time required to tune a loop be minimized. The auto-tuning user interface will be used by a variety of people in the plant and thus must be designed to be intuitive and easy to use.

The manufacturer may provide multiple ways to launch this auto-tuning capability. For example, the tool may be launched from the Start button, as an option in the function block view, or from the PID faceplate provided in the operator display.

12.3.1   Auto-tuning Application

When the auto-tuning tool is launched in context, the "auto-tune" interface for the selected PID block is displayed as shown in Figure 12-3.

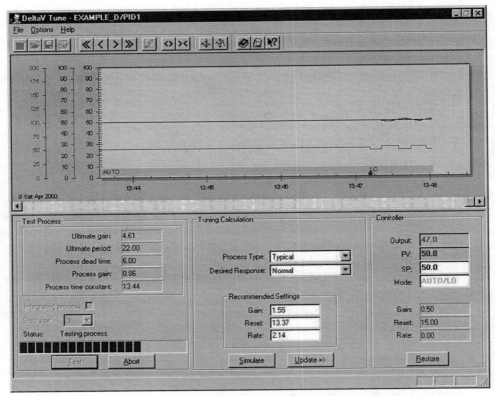

Figure 12-3. Auto-Tune Interface

The on-line values of the controlled parameter PV (top line in plot varying about the SP Value), the setpoint SP (top straight plot line), and the control output of the PID block (lower line in plots) are shown in the right portion of the screen. Also, a trend of these parameters is automatically displayed. The current tuning of the PID is also shown in the right portion of the screen: proportional gain, reset (integral gain), and rate (derivative gain). To initiate the PID tuning process, the Test button must be selected. Once the Test button has been selected, the output of the PID is changed automatically from its initial value by the default step size shown immediately above the Test button. The default step size of 3% is appropriate for most processes. However, if the controlled parameter measurement contains signification process noise, then a larger step size may be selected. By increasing the step size, the resulting change in the controlled parameter may be easier to detect accurately. If the process gain is quite large, a smaller step change may be warranted to avoid the process upset often associated with a large change in a controlled parameter.

After the Test button has been selected and the initial step change in the manipulated parameter has been specified, no further user input is required. Once a response is detected by the auto-tuning application, the manipulated parameter is changed in the opposite direction from its starting direction. After this second change in the manipulated parameter, no further action is taken until the controlled parameter in response to the manipulated parameter change passes through its value at the start of the test. Each time the control parameter transitions through its value at the start of the test, the manipulated parameter is automatically changed by the step size in the opposite direction from the last change made in the manipulated parameter. This method for determining the process gain and dynamics is known as the relay oscillation technique. [1] The period of oscillation established by the control output switching is known as the ultimate period. Based on the step size and the amplitude of the changes in the controlled parameter, it is possible for the auto-tuning application to calculate the ultimate gain.

>**Ultimate gain** – Proportional gain required to achieve sustained oscillation of the controlled and manipulated parameter (with integral and derivative gain set to zero).

>**Ultimate period** – Period of oscillation achieved for sustained oscillation when the proportional gain is set to the ultimate gain (with integral and derivative gain set to zero).

Based on the ultimate gain, the ultimate period and the time required to see an initial response to a step change in the manipulated input, it is possible for the auto-tuning application to calculate the process gain, time constant, and deadtime. These parameters, as well as the ultimate gain and ultimate period, are shown in the left pane of the auto-tuner interface.

The ultimate gain and period can be determined manually by setting the integral and derivative gain to zero then, with the PID in Automatic, gradually increasing the proportional gain in steps, changing the setpoint, and observing the response to the setpoint change. This procedure is repeated until sustained oscillation is observed. However, with a slow-responding process this manual technique is very time consuming. The relay oscillation technique allows the process gain and dynamics to be quickly determined through automated testing. Also, the changes in the process input are known and thus can be selected to minimize the impact of testing on process variation.

Based on the process gain and dynamics identified through the automated test, the auto-tuning application can calculate a recommended setting for the PID proportional, integral, and derivative gain. By selecting the Update button, these recommended tuning settings may then be transferred to the PID block in the controller. To use the tuning determined by the application, the user must just select Update and the tuning values will be automatically written to the PID block.

In summary, when such a tool is available, to tune most of the control loops found in the process industry the user need only select Test, wait for the test to complete and then select Update. So, with two clicks of the mouse, a control loop may be tuned.

To provide some user flexibility in specifying the control loop response, the user may select the tuning rule used to calculate the recommending tuning with the identified process response. This option is provided in the Tuning Calculation area of the interface through the selection of Process Type and Desired Response. Using the Process Type selection, the user may choose Typical PI or PID or select Pressure Loop, Temperature Loop, or Flow Loop. The recommended tuning is calculated based on the loop type selection and the identified process gain and dynamics. The tuning of a pressure loop would, for example, be different than that of a flow loop, even though each has the same identified process dynamics and gain. In this manner, the tuning may be matched to the process type. If the process type is left at the default setting of Typical PI, then the tuning is calculated to give a typical response that would be good for most processes. Similarly, the Speed of Response selection allows the user to choose a slower or faster response to a change in setpoint and load disturbance.

12.3.2   Simulation of Response

One of the useful features supported by the auto-tuning application described above is that the user may examine the expected closed loop response before transferring the recommended tuning to the control loop. To view the expected closed loop response, the Simulate selection should be chosen after testing is complete. In response, a screen will be displayed that includes a plot in the lower portion of the display that shows the closed loop response for a setpoint change, as illustrated in Figure 12-4.

By default, the closed loop response for a step change in set point is displayed. As illustrated in Figure 12-4, the change in the manipulated parameter and its impact on the controlled parameter is shown for the recommended tuning. By selecting Disturbance in the plot options, the closed

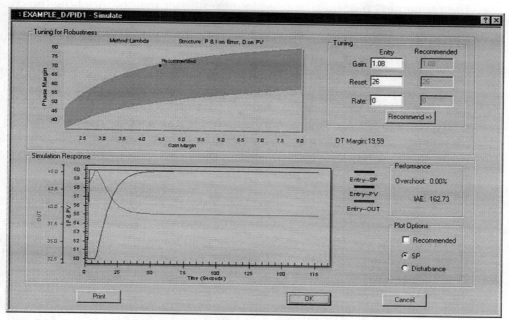

Figure 12-4. Simulation of Loop Response

loop response for a step change in an unmeasured process disturbance will be displayed. If the process has been correctly identified during testing, the calculated closed loop response should accurately reflect the actual closed loop response when the recommended tuning is transferred to the control loop. This capability may be used to examine the impact of the recommended tuning on the closed loop response.

In the upper portion of the Simulate screen is a plot that shows the robustness of the closed loop control for recommended tuning [2].

> **Control Robustness** – A measure of how well control copes with differences between the true process gain and dynamics and that assumed in setting PID tuning.

This robustness plot shows the gain margin and phase margin for the recommended tuning. Also, the acceptable region of operation is shown by the dark band in the top plot.

> **Gain margin** – Change in process gain that would be required for an unstable closed loop response.

> **Phase margin** – Added phase shift through the process that would be required for an unstable closed loop response.

The robustness plot may be used to gain further insight into how the control loop will perform with the recommended tuning when the process gain or dynamics change. The gain margin gives a direct indication of how much the process gain may change before the control loop will go unstable. For example, if the recommended tuning is shown to have a gain margin of 3, then when this tuning is applied, stable control will be observed even though the process gain changes by a factor of three. If you said, "I really want to have a 5-gain margin," that is, the process gain could change by a factor of five and you would still have stable control, by clicking in the dark band on the robustness plot for a gain margin of 5, you would see the tuning that provides this gain margin along with a plot of the response.

Selecting an operation point in the left portion of the robustness plot provides higher proportional gain since the gain margin is smaller. Selecting a point in the lower portion of the robustness plot will provide tuning with lower phase margin. That means that if the process changes by more than the gain margin then an unstable response may be observed. Selecting a point associated with greater gain margin or phase margin will make the control loop performance less susceptible to changes in process gain or dynamics.

The impact of a selection on the robustness plot is reflected in the plot of the response to a setpoint change. So by making different selections in the robustness plot, it is possible to choose a tuning that gives a desired response with a known gain and phase margin. Tuning determined in this manner can be transferred to the recommended tuning through the selections provided in the upper portion of the screen. The robustness plot can be a very effective tool for not only evaluating the recommended tuning but also for refining the recommended tuning to meet specific requirements.

12.4 Commissioning – Sticky Valves and Other Field Challenges

The long-term performance of a plant is directly impacted by the care with which the control system is commissioned. Establishing control loop tuning is an important part of the work needed to commission a new plant or an area of the plant in which measurement or control changes have been made. Such work can be rewarding since the impact of a change made in control setup and tuning is immediately seen in the plant operation. However, loop tuning can, at times, also be challenging because of problems in establishing tuning and stable operation of the plant when there is a failure or degraded performance of field devices. To resolve these problems, some investigation and often a trip to the field are required to observe valve movement or the physical condition of a field device and its associated wiring installation. Even when commissioning a new installation, it is common to encounter many problems. A person who does this work quickly learns some valuable lessons about commissioning:

Lesson 1 – Be aware of problems in field installation.

Many of the problems associated with control system commissioning have nothing to do with the setup or tuning of the PIDs.

It is important to be aware of problems in field installation that may prevent a control loop from working correctly. For example, the first step in loop commissioning should be to verify that the transmitters and final control elements are connected to the correct inputs and outputs in the control system. Also, if an isolator has been installed between a four-wire transmitter and the control system, it is important to verify that the inputs to the control system have not been reversed. If the input has only been verified at mid-range, then it is possible that such a reversal in wiring may have gone undetected during wiring checkout that is normally done in the initial phase of commissioning. When the problem is with a valve or actuator, the process input may be reversed because of the I/P setup and the actuator Fail Safe setup. Such reversals in the process input should be accounted for in the control system through the Increase Open/Close setting of the Analog Output block. These problems can be quickly detected by field verifying the value of 0% and 100% controller output.

Lesson 2 – Observe loop response.

Establish an understanding of the process gain and dynamics by trending the control block PV and OUT using trend screens that are standard in most control systems and by introducing a small step change in the manipulated parameter while the loop is in Manual mode.

In some cases, when a step change is made in the manipulated parameter, no change may be observed in the controlled parameter. If a loop shows this behavior, then there is a good chance that it is caused by:

1. A mistake in the wiring or configuration associated with the transmitter and/or final control element input or output. Solution: Correct wiring or configuration.

2. The valve or actuator is sticking, that is, the valve did not respond to a change in the output current signal. Solution: depending on the cause of sticking, repair valve or actuator, or install a valve positioner.

3.   The transmitter is broken or incorrectly installed. Solution: Repair transmitter and/or correct installation.

Lesson 3 – Validate loop setup.

It is never a good idea to just place a control loop in automatic mode using the default tuning and setup. The positive feedback caused by any unaccounted-for reversals in the transmitter and control system output can cause the manipulated valve, damper drive, variable speed drive to quickly go to a full open or closed position. In an operating plant, such a mistake can cause a process disruption or even a plant shutdown. Whether the PID option for Direct acting and the AO block option for Increase to Close are required can be quickly determined by changing the PID output in manual mode and observing the response.

Since the valve or actuator is often the least reliable component in a control loop, a sticky valve or loose mechanical connection associated with a damper drive may be the cause of no response to a change in the manipulated parameter. With a sticky valve, as the current input to the valve actuator changes, the valve may remain stuck at its last position. A significant force may be required to change the valve position. This is especially true with sliding stem valves, which exhibit friction between the packing and the valve stem. Loose connections associated with a damper drive may mean that changes by the damper drive are not reflected in the position of the manipulated damper.

When a valve positioner has not been installed on a valve, the control system may interface to the actuator through an I/P transducer. In the case of a sticky valve, the pressure change provided by the I/P may not initially be enough to cause the valve stem position to change, and multiple changes in the control system output, and to the pressure delivered to the actuator, may be required to move the stem. Once the stem begins to move, less force will be required, and the actuator pressure may now exceed the pressure that is needed to achieve the desired position. As a result, a larger than intended change in flow may be observed. This nonlinear response associated with a valve actuator or damper drive, that is, the controlled parameter does not reflect changes in the input to the manipulated parameter, can significantly degrade control performance and make it difficult to commission the control system.

When a valve positioner is installed on the actuator, as long as it is functioning properly it will automatically increase the actuator pressure until the valve moves to a position that matches that requested by the control system. Feedback of the stem position is used by the positioner to sense if the valve has responded. Thus, it is possible for the valve position to eliminate overshoot or offset.

When a control loop is placed in automatic control, it is easy to detect if a valve or damper is not responding to the control system by observing the response of the controlled parameter to control system changes in the PID output. An example of the response that may be observed with a sticky valve is shown in Figure 12-5.

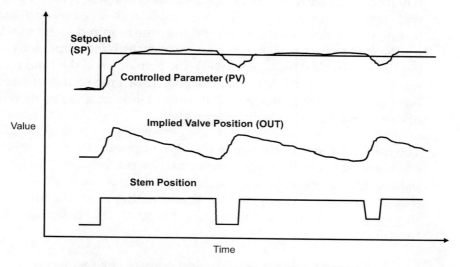

Figure 12-5. Impact of Sticky Valve on Automatic Control

As illustrated in Figure 12-5, in the absence of a valve positioner a significant change in the setpoint may cause a change in the control system output, as shown by the PID output, that is large enough to generate the actuator pressure needed to move the valve stem but result in the valve overshooting the target position and introduce cycling in the control. Small output changes made by the control system for small deviations in the controlled parameter from setpoint must accumulate before enough force is developed to move the valve. When the valve moves in response to accumulated changes in PID output, then a large change in the controlled parameter may be observed. The change in the control system output in response to this large change in the controlled parameter may be enough to again cause the valve to move. Thus, a sustained, characteristic cyclic pattern will develop in the controlled parameter and the manipulated parameter. The amplitude and period of the cyclic pattern will vary depending on the controller tuning and the force needed to cause the valve to move. The cyclic behavior caused by a sticky valve cannot be eliminated through tuning. Changes in tuning will only impact the period of the cycle that develops. The only way to eliminate this type of behavior is to install a valve positioner.

Cyclic behavior in automatic control may also be observed if deadband is introduced by a mechanical linkage between the actuator and the final control element. However, an important difference is that the cyclic pattern caused by deadband may die down over time. Deadband is characterized by the final control element responding in a linear fashion to small control system output changes that are made in one direction. However, when changes are made in the opposite direction, the final control element will respond only after a sufficient amount of change has been made in this new direction. This may extend the time required to get to setpoint or for the PID to compensate for changes in disturbance inputs. By examining the mechanical linkage between the actuator and the final control element, it is often possible to identify and correct the source of deadband. For example, wear in the pins that connect linkage arms is often the cause of deadband and can easily be corrected.

The minimum size of change in the PID output that the control element can respond to can be quickly confirmed by placing a control loop in Manual and then adjusting the control system output in one direction in small steps, for example, changes of 0.1%, and noting when the final control element begins to move. By this means, it is possible to determine the output resolution.

> **Output Resolution** – Smallest percent change in control system output that will be seen in the controlled parameter value.

After a series of changes have been made in one direction, it is possible to determine the deadband by making a series of changes in the opposite direction.

> **Output Deadband**– Smallest percent change in direction of control system output that is required to change the value of the controlled parameter.

If a measurement of the position of the final control element is an input to the control system, then this input may be used to confirm that the final control element has responded to the control system output. When such a measurement is not available, movement of the final control element to moderate changes in the PID output may be visually confirmed by going to the field and observing the final control element response to a change in the control system output. Also, by trending the control parameter on an expanded scale along with the PID output and making small step changes in the PID output, it is possible to confirm the changes in the PID output impacted the control parameter.

The most common problems in commissioning a control system often can be traced to the fact that a positioner has not been provided with the valve, or the positioner provided with the valve has not been properly installed or has malfunctioned. The rule of thumb is that to achieve best control performance, all regulating valves should be equipped with a positioner. Without a positioner, the control performance that may be achieved is very limited when a valve is sticking—which, as we have seen, is inherent in most valves. Thus, when you are commissioning a control system, it is a good idea to establish up front which valves have positioners, and to clearly communicate the expected limitations in control performance for valves not equipped with positioners. If some of the final control elements are dampers, then it is wise to inspect the actuator linkage to determine if loose tolerances in linkage joints could be a source of deadband.

An installation inspection may show that positioners are installed on all valves and that linkages associated with dampers and other final control elements are in good condition. Even so, it is always a good practice to begin loop tuning by placing the loop in Manual, making some small changes in the control loop output, and then observing to see if the process responds to these changes. When no response is observed, then an effort should be made to investigate and address any physical or electrical problems before attempting to tune the control loop.

12.5 Characterizing Loop Gain

During the commissioning of a control loop, the process response should be observed over a wide range of operation. The process gain of a self-regulating process may be determined based on the steady-state value of controlled parameters at specific manipulated input values. By plotting the steady-state controlled parameter value versus the manipulated input value, it is easy to determine if the process gain is constant over the operating range. The steady-state controlled parameter value of a self-regulating process may be determined by placing the control loop in Manual and then maintaining the manipulated parameter at a constant value.

As we have seen, the steady-state value of the controlled parameter is the value observed after the process has fully responded to a change in the manipulated parameter. For example, if the controller output to a valve normally varies between 30 and 50% during normal plant operations, a steady-state value of the controlled parameter might be observed when the valve is maintained at points over this range, for example, 30, 35, 40, 45, 50%. If the plot of the controlled parameter versus the valve position for steady-state conditions is a straight line, then the process gain is constant. To determine the gain of an integrating process over a wide operating range, it is necessary to plot the rate of change in the controlled parameter versus the process input.

From a control perspective, it is highly desirable that the process gain be constant. If the process gain is constant, then the same proportional gain may be used over the entire operating range of the control loop. However, if for example the valve characteristic has not been selected based on the process requirements, then the installed characteristic could be non-linear as illustrated in Figure 12-6.

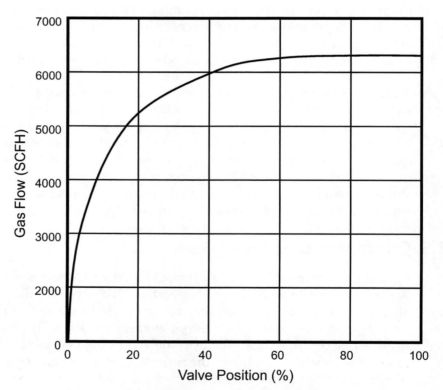

Figure 12-6. Example of Non-linear Installed Characteristics

From this plot, the process gain may be determined at any operating point by dividing the per cent change in the controlled parameter by the per cent change in the manipulated parameter. As illustrated in this example, the process gain varies from 0.5 to 4, that is, the process gain changes by a factor of eight. Even if the loop tuning is selected to have a high gain margin, such a large range in process gain can cause slow response in the low gain region and potentially unstable operation in the high gain region. If the process gain is known to change over the operating range of the final control element, then loop tuning should be established in the high gain region to ensure that stable operation will be achieved over the entire operating region.

From a control perspective, the product of controller gain and process gain should be a constant value over the entire operating region of the final control element. If the process gain is changing, then the impact is the same as changing the control loop gain of a process with constant gain. Thus, changes in process gain can have a large impact on control performance.

If during the process of tuning a control loop it is discovered that the process gain changes significantly over the normal operating range of the final control element, then the question becomes, "How do I correct for significant changes in the process gain?" Changing the valve characteristics to provide a linear gain over the operating region may not be an option when working with a valve because of the time and expense associated with changing valve characteristics. To ensure stable operation, the tuning may be established at the operating point of maximum process gain. However, in the low gain region, control performance will be sluggish. When you are confronted with this situation, a better solution is to compensate for the changes in process gain by installing a characterizer block between the PID and Analog output blocks as illustrated in Figure 12-7. If the control system does not support this block, then often it is possible to construct equivalent capability using other tools of the system provided for doing calculations.

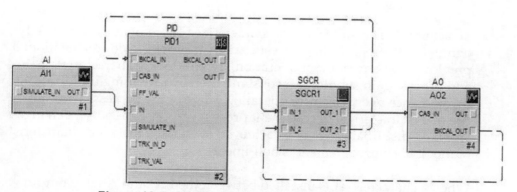

Figure 12-7. Signal Characterization in Control Path

Characterizer blocks are available in many Foundation Fieldbus devices and in some control systems. The relationship between the primary inputs and the output of the characterizer block may be defined by 21 x,y pairs over the final control element operating range. Input values that fall between these points are automatically determined by the characterizer block using linear interpolation. The objective is to define the characterizer relationship such that the product of the characterizer gain and process gain is constant over the entire operating range of the final control element. The curve defined by the characterizer points appears as the inverse of the plot of the final control element installed characteristic, as illustrated in Figure 12-8.

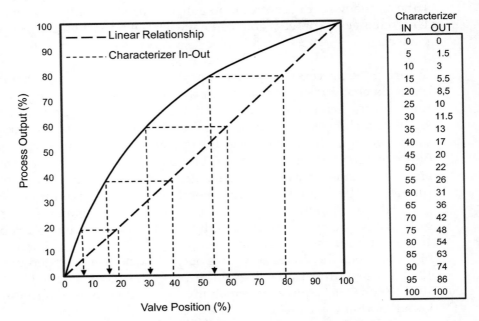

Figure 12-8. Example of Characterizer Setup

To achieve best results using a characterizer, more points are normally defined in the regions of greatest change in process gain. As mentioned previously, at any point of operation, the product of characterizer gain and process gain should be a constant, thereby providing uniform control performance over the entire operating range of the process. Through the use of a characterizer, a consistent process response to setpoint changes and load disturbances may be achieved even if the installed characteristic of the final control element is non-linear.

When a characterizer is installed between the PID and Analog output blocks, the back calculation connection must also go through the characterizer to ensure bumpless transfer and communication of status information needed to avoid PID windup. The characterizer block provides a second set of connections for the back calculation path. To provide the PID back calculation value needed for bumpless transfer, the characterizer block essentially does the inverse of the calculation done in the forward path of the characterizer block. This capability is a standard feature of the Foundation Fieldbus characterizer block available in some field devices and control systems. The only restriction in supporting this back calculation path is that the characterization points be defined as increasing or decreasing in a monotonic manner.

> **Monotonic** - Either entirely *nonincreasing* or *nondecreasing*.

Since processes are designed to provide this type of behavior, this restriction in defining plot points does not limit the application of the characterizer in control applications.

12.6 Pairing of Parameters, Decoupling

In Chapter 9, Process Characterization, the examples focused on single input/single output and single input/multiple output processes. However, in some cases, the process equipment is characterized as having multiple manipulated inputs. This type of equipment configuration is often referred to as a multiple input/multiple output process. The process control consists of multiple PID blocks where a process output and process input pair are associated with each PID. The challenge in designing and commissioning controls for this type of process is interaction, that is, each manipulated input may impact multiple controlled parameters. The question then becomes, "Which manipulated input should I adjust to maintain a controlled parameter at setpoint and what is the relative importance of the control parameters in achieving the operating objectives?" The answer to this question may not be obvious. Adjusting a manipulated parameter to maintain one controlled parameter at setpoint may also impact another controlled parameter. This deviation in the second controlled parameter may in turn require a second manipulated parameter to be changed, which may then impact the first controlled parameter. In general, this interaction may impact the control of two or more loops. A cycle may be established in which the loops fight each other. In such cases, it is important to pair the controlled and manipulated parameters to minimize the interaction between the control loops. Such interaction can significantly degrade control and, in some cases, can cause sustained oscillation.

> **Pairing of control loops** – Selection of controlled and manipulated parameters to minimize interaction between control loops.

To determine the correct pairing of controlled and manipulated parameters, it is often necessary to do a series test in which a step change is introduced into each manipulated parameter and the impact on each controlled parameter is observed. Based on this test, it is possible to calculate the process gain associated with each input and output combination. In most cases, pairing the controlled parameter with the manipulated parameter that maximizes the process gain will give best control performance and will minimize loop interaction.

If the process gain associated with a single controlled parameter and multiple process inputs is nearly the same, then significant loop interaction will be experienced independent of the pairing of controlled and manipulated parameters. This type of process is often referred to as being highly interactive.

During system commissioning, the degree of interaction associated with two or more loops may not be immediately obvious. For example, during startup, all loops may be initially placed in Manual. When one loop is commissioned, then no interaction will be observed since the interacting loops are in Manual. Only when you are attempting to tune a second loop with the first loop in Automatic will the interaction between these loops become obvious.

The textbook solution for addressing interactive processes is to install a de-coupler between the PID and Analog output blocks as illustrated in Figure 12-9. A change in the output of each PID may then impact one or more manipulated parameters with the net effect being that only one controlled parameter is impacted by the change.

In theory, a decoupling network may be used to address interactive processes. However, in practice decoupling networks are seldom used because of the complexity involved in commissioning and maintaining the proper values of the gains used in the decoupling network. In actual practice, the fighting between interactive loops is most often addressed by simply detuning one of the control loops. The valve associated with the detuned loop will change very slowly. Thus, the two loops will tend not to interact. When taking this approach it is necessary to select the controlled parameter that should be tightly controlled. The other interacting loop may be loosely controlled since the gain of the de-tuned PID may be a factor of three or four smaller than what is normally applied to a non-interacting loop. By making the loop response slow, the interaction may be broken.

If there is an operational requirement for both controlled parameters to be tightly controlled, or if multiple control loops are found to interact, it is necessary to take another approach. As will be discussed in a later chapter, one approach to effective control of interactive loops is the use of model predictive control. The structure of model predictive control addresses process interaction without making compromises in control performance.

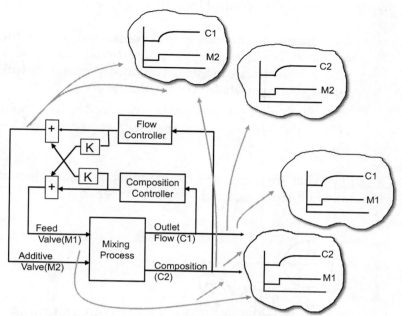

Figure 12-9. Decoupling Process Interaction

12.7 Workshop – PID Tuning

This PID Tuning workshop provides several exercises that may be used to further explore loop tuning. The heater process shown in Figure 12-10 will be used in this workshop. See the Appendix for directions on using the web-based interface.

Step 1: Using the web-based interface, open the PID Tuning workshop in the Workspace tab to view control and simulation of the heater shown in Figure 12-11. Try manually tuning the PID in this example using manual tuning techniques:

- With the loop in Manual, change the manipulated parameter by a step to determine the process deadtime and time constant.

- Verify that the default control action is correct. Set the reset time equal to the deadtime plus the time constant.

- Set the proportional gain to a conservative value, for example 0.1, and place the loop in automatic control.

- Change the setpoint and observe the control response. Gradually increase the proportional gain (leaving the reset at the value previously entered) to achieve the desired closed loop response.

<u>Step 2</u>: Try tuning the control loop for changes in disturbance inputs:

- Introduce a disturbance in the process and observe the control response. Try increasing the proportional gain by 50% and introduce a change in the disturbance. Was the control able to more quickly compensate for the disturbance?

- With the higher proportional gain, introduce a setpoint change. How did the increase gain impact the response to a setpoint change?

<u>Question</u>: What are the trade-offs that must be made when tuning control for both load disturbance and setpoint changes?

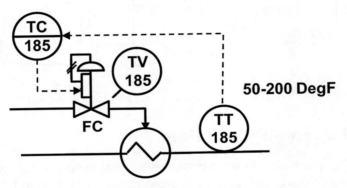

Figure 12-10. Process for PID Tuning Workshop

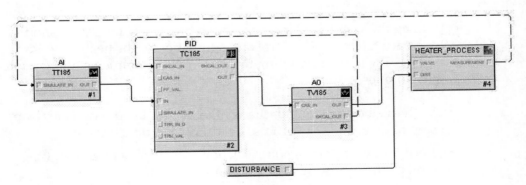

Figure 12-11. PID Tuning Workshop Module

12.8 Workshop Discussion

<u>Step 1</u>: Were the default values for direct/reverse control action and increase/decrease to open setup correct for this control loop? After manually setting the reset, was it possible to achieve the desired closed loop response just by adjusting the proportional gain?

<u>Answer</u>: Yes, in this example the PID default settings should be reverse action and increase to open. These are the most frequently used values. No adjustment of reset should be made once it is set based on the process dynamics.

<u>Step 2:</u> When you increase the proportional gain, did you see an improvement in the response to changes in the disturbance input?

<u>Answer</u>: When the proportional gain was increased, it should have improved response to changes in the disturbance input. However, the response to setpoint changes with the higher proportional gain may have been too responsive and may have shown oscillatory response. Thus, when tuning for changes in both setpoint and disturbance inputs, then it is often necessary to set proportional gain to provide satisfactory response for both types of changes.

References

1.   Åström, K. and Hägglund, T. *Automatic Tuning of PID Controllers.* Research Triangle Park: ISA, 1988 (ISBN: 978-1-55617-081-2).

2.   Blevins, Terrence L., McMillan, Gregory K., Wojsznis, Willy K. and Brown, Michael W. *Advanced Control Unleashed: Plant Performance Management for Optimum Benefit*, ISA, 2003 (ISBN: 978-1-55617-815-3).

13

Multi-loop Control

Single-loop PID is the most common control technique used within the process industry. However, in many cases, control performance may be improved through the use of multi-loop control techniques. While most modern control systems fully support the configuration, checkout and commissioning of multi-loop control, these more complex control techniques may require added time to configure, check out, and commission. Also, additional training is needed to commission and maintain multi-loop control strategies. Thus, it may be difficult to justify multi-loop control in a new installation if single-loop control is sufficient to allow the plant to come on-line. However, once the plant is on-line, or in a plant that is already running, the reduced process variation and other improvements that are possible through the use of multi-loop control may justify the cost, effort, and added measurements required to implement multi-loop control. This chapter addresses the multi-loop control techniques that are most commonly used in the process industry.

13.1 Feedforward Control

One of the fundamental limitations of single-loop control is the fact that all control action is based on the error between the setpoint and the controlled parameter. If there is no error, then the control output will remain at its last value. To correct for process disturbance input using single-loop control, the disturbance input must impact the controlled parameter. Only *after* a change is seen in the controlled parameter will the manipulated parameter be adjusted to correct for the disturbance. However, if a measurement of the disturbance is available to the control system, then through the application of feedforward control, it is possible to anticipate the impact of the disturbance. Measured disturbances may be fed forward to the control system so that action can automatically be taken to correct

247

for the measured disturbance before it impacts the controlled parameter. The implementation of feedforward control is always done in conjunction with feedback control, as shown in Figure 13-1.

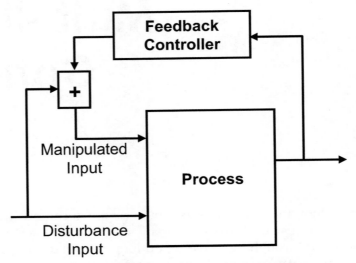

Figure 13-1. Basis of Feedforward Control

A key point when considering the use of feedforward control is that the process disturbance must be a measured input to the control system. If the disturbance input is now unmeasured, then to apply feedforward control a means to measure the disturbance must be added as a control system input. In a basic feedforward strategy, a scaled value of the measured disturbance input is simply added to the feedback controller contribution to determine the manipulated parameter value. The feedforward contribution associated with a change in a disturbance input will cause the manipulated parameter to immediately change.

In practice, as mentioned above feedforward control is always applied in combination with feedback control because there are always some disturbances that aren't measured, and for those, the feedback is still required. Also, the identified process gain and dynamics on which feedforward control action is based are never perfect.

Feedforward control in combination with feedback control is an example of MISO control, that is, a multiple input, single output control strategy.

> **MISO control** – Control based on the use of multiple process inputs and outputs to determine a manipulated process input.

Through the use of feedforward control it is possible to reduce variation in the controlled parameter when a process is subjected to significant process disturbances.

13.1.1 Dynamic Compensation

When you are implementing feedforward control, the difference in the process response to a change in the manipulated parameter versus a change in a disturbance input should be known. In some cases, the response of the controlled parameter to a change in a measured distur- bance input exhibits the same time constant and deadtime as its response to a change in the manipulated parameter. If these responses are the same, then the scaled feedforward input may be directly added to the feedback contribution as previously shown. However, in general, these responses differ. For example, when a disturbance input to a process changes, it may take five seconds for this change to impact the controlled parameter. When the manipulated input changes, it may only take two seconds for this change to impact the controlled parameter. To correctly compensate for a change in a measured disturbance input, this dynamic difference must be taken into account in the feedforward implementation.

The response of the controlled parameter to a step change in the measured disturbance (with all other process inputs maintained constant) can often be characterized as first order plus deadtime. Based on knowing the response of the controlled parameter to changes in measured disturbance and manipulated input, it is possible to dynamically compensate for any differences in time constant and deadtime.

> **Dynamic compensation network** – Network included in the feedforward path to account for differences between the process response to changes in a disturbance input and changes in the manipulated input.

If the process response to changes in the manipulated and disturbance inputs can be characterized as a first order plus deadtime, then the difference in the process response may be compensated for by using a deadtime block and a lead-lag block in the feedforward path as shown in Figure 13-2.

Not surprisingly, the parameters associated with the lead/lag and dead- time blocks must be set based on the process response to changes in the manipulated input and the measured disturbance input. Thus, to apply dynamic compensation, it is necessary to know those two process responses.

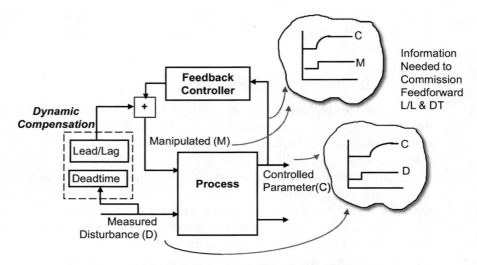

Figure 13-2. Feedforward Dynamic Compensation

As previously discussed, the process response to a change in a manipulated input may be determined by switching the feedback control to Manual and making a step change in the manipulated input. If the disturbance input may be manipulated, the same procedure may be followed. However, it may not be possible to make the disturbance input change by a step. For example, the disturbance input to a process may be outside air temperature. In such a case, it may be necessary to examine historical data to determine how changes in the disturbance input have impacted the controlled parameter. Because of the difficulty in determining process response to a change in the disturbance input, it may not be easy to determine if dynamic compensation is required or what values should be used in the lead/lag and deadtime blocks. For that reason, feedforward control may have to be implemented without dynamic compensation or perhaps should not be implemented at all.

When feedforward control may be implemented, the PID function blocks in Foundation Fieldbus devices and in many control systems support a feedforward input. The scaling of the input and addition of this input to the feedback contribution is done internally within the PID block. Dynamic compensation may be added using a deadtime block and lead/lag block in the feedforward path as shown in Figure 13-3.

The lead/lag block lead time constant and lag time constant are determined by the LEAD_TIME and LAG_TIME parameter. The deadtime introduced by the deadtime block is determined by the DEAD_TIME parameter. Also, the PID block provides a parameter for feedforward gain, FF_GAIN, that is used to scale the feedforward input. The characterized process responses to changes in the manipulated and disturbance

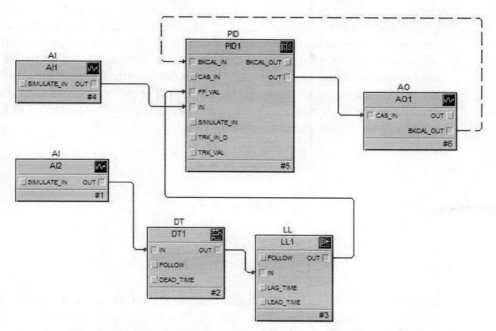

Figure 13-3. Dynamic Compensation in Feedforward Path

inputs are used to set the lead/lag block parameters and the feedforward gain in the PID block as illustrated in Figure 13-4.

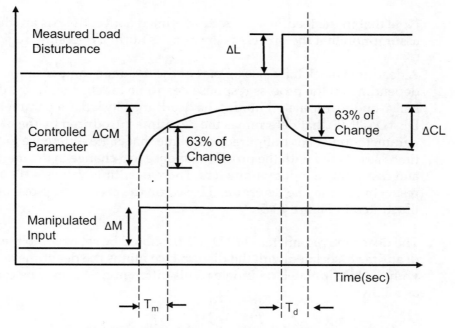

Figure 13-4. Step Test to Determine Process Dynamics

Based on the characterized responses, the lead time should be set equal to the time constant (presuming a first order plus deadtime process) associated with the process response to a change in the manipulated input. The lag time should be set equal to the time constant associated with the process response to a change in the disturbance input. The feedforward gain should be set equal to the negative of the ratio of the process gain for a change in the disturbance input to the process gain for a change in the manipulated input as summarized below.

$$FF_{lead} = T_m$$

$$FF_{lag} = T_d$$

$$FF_{gain} = -\left(\frac{Gain_{disturbance\ input}}{Gain_{manipulated\ input}}\right)$$

If the time constants are the same for both a load disturbance and a change in the manipulated parameter, then the lead/lag block does not need to be included in the dynamic compensation network.

In the feedforward gain calculation, the actual process gain value is used, rather than its absolute value. As a result, the feedforward gain may be positive or negative depending on the gains associated with changes in the load disturbance and manipulated parameter.

Note that in the feedforward setup outlined above, there is an implicit assumption that the process is linear in its normal operating region.

A deadtime block may also be part of the dynamic compensation network depending on the process response deadtime for changes in the disturbance and manipulated input. The deadtime block has a parameter, DEAD_TIME, that determines the deadtime introduced by the block. This parameter must be configured to a value that is the difference in the deadtimes associated with the process response to a change in the manipulated and disturbance input parameters. These deadtime values, DT1 and DT2, respectively, may be determined based on the process step response as illustrated in Figure 13-5.

The deadtime parameter DEAD_TIME should be set equal to the deadtime associated with the disturbance minus the deadtime associated with the manipulated parameter. This difference, FF_{dt}, may be calculated as follows:

$$FF_{dt} = DT_2 - DT_1$$

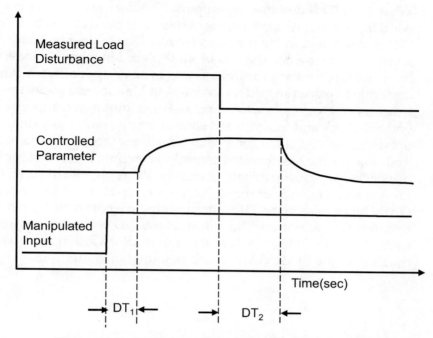

Figure 13-5. Step Test to Determine Process Deadtime

When commissioning a feedforward network, a question that may come up is, "What if the deadtime associated with a manipulated parameter is longer than the deadtime associated with a change in the load disturbance?" A negative deadtime implies that a change in the manipulated parameter must be made on some future value of the load disturbance. Since it is not possible for the deadtime block to predict the future, the deadtime parameter can only be set to positive values. So, when the deadtime is calculated to be a negative number, deadtime should not be included in the dynamic compensation network. Also, if the calculated deadtime is zero or small in value compared to the process time constant, then the deadtime block may provide little benefit in the compensation network. The best that can be done for these types of processes is to immediately act on the load disturbance value.

13.1.2 Alternate Implementations

There are a couple of alternate implementations of feedforward control that are supported by a control system that uses Foundation Fieldbus blocks. These alternate implementations are not commonly used but can be useful in meeting special requirements. A brief description of these alternate implementations is given in this section.

When the PID is designed to support a feedforward input, the feedforward input after dynamic compensation is applied can be connected to the PID as illustrated in the previous section. One of the ramifications of this approach is that when the PID block is placed in manual mode, both feedback and feedforward control are lost. In some special cases, it may be desirable to place the feedback control in Manual and allow the feedforward control to continue to function. For example, in composition control, both feedback and feedforward control may be required to maintain product specifications. If the composition analyzer were to fail then it might be desirable for the feedforward control to continue correcting for load disturbances. To achieve this functionality, the feedforward control must be implemented outside of the PID. One way to do this is to use the BIAS/ GAIN block to sum the PID output with the feedforward input after dynamic compensation. The output of the PID becomes the cascade input, CAS_IN, of the BIAS/GAIN block. The feedforward input is connected to the IN_1 of the BIAS/GAIN block as shown in Figure 13-6.

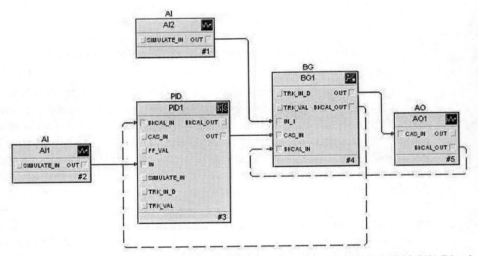

Figure 13-6. Alternate Feedforward Implementation – Using BIAS/GAIN Block

The PID output connected to the CAS_IN of the BIAS/GAIN block is used as a bias value that is added to the feedforward signal connected to IN_1. The feedback contribution is determined by the PID output scaling and output limits. Scaling of the feedforward input is accomplished through the GAIN parameter of the BIAS/GAIN block.

If the feedforward summing function is implemented outside of the PID, the PID output should be scaled (using OUT_SCALE) to allow both positive and negative contributions, for example, the OUT_SCALE range would be configured to be -100 to +100. This scaling determines the range of adjustment that can be made by the feedback component. For example, consider the case where the disturbance input is scaled 0–100%. If the PID output is scaled -100 to +100 then the feedback could completely override any contribution by the feedforward input. If the PID output contributed a bias of -100 then the BIAS/GAIN output can be driven to zero independent of the feedforward input. Similarly, if the feedforward signal is zero, then by the PID contributing a bias value of +100, the BIAS/GAIN output will be driven to 100%. If the span of the PID output has been set in this manner, then the PID output should gravitate toward 50% of range when the PV is at SP for normal operating conditions, that is, at 50% of range the feedback contribution is zero since output scaling is -100 to +100.

When you are using the BIAS/GAIN block to build the feedforward summing function it is necessary to consider the impact of scaling the PID output and the gain of the BIAS/GAIN block. However, this approach provides a way for feedforward control to function even when the PID is placed in Manual. To ensure correct PID initialization to provide bumpless transfer, the BIAS/GAIN control option Act on Initialization Request (IR) as illustrated in Figure 13-7 must be selected. Enabling this option causes the bias to be automatically initialized to provide bumpless transfer when the mode of the AO block is changed to Cascade.

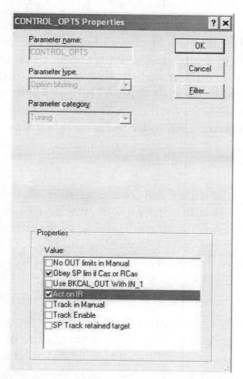

Figure 13-7. External Feedforward – BIAS/GAIN Setup

In certain cases, a process measurement such as the feed flow to the process may reflect a change in the gain relationship between the controlled parameter and the manipulated parameter. Such a change in the process may be compensated for through the use of multiplying feedforward. The RATIO block may be used to build the feedforward multiplier outside of the PID as illustrated in Figure 13-8.

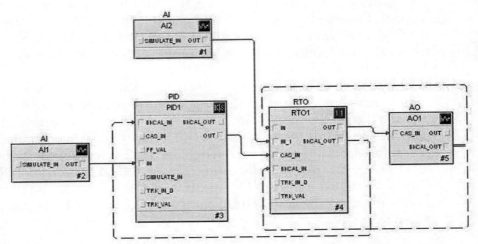

Figure 13-8. Alternate Implementation – External Feedforward – Ratio

Similar to the alternate implementation of the feedforward summing function, the PID output is the cascade input, CAS_IN, of the RATIO block. The scaled PID output and the disturbance input provided through IN_1 are multiplied to provide the RATIO block output. The PID output should typically be scaled over a narrow range, for example, 0.5 to 1.5, allowing the feedforward signal to be reduced by 50% or increased by as much as 50%. The PID output scaling should be configured based on the desired range of feedback adjustment.

For most applications, the feedforward control capability incorporated in the PID block is sufficient. However, where there is a clear requirement for an external summing or multiplying of feedforward input, then it is possible to provide this capability using the BIAS/GAIN or RATIO block in conjunction with the PID.

13.1.3 Workshop – Feedforward Control

In this feedforward control workshop, the process example is a heater shown in Figure 13-9. The outlet temperature of the heater is controlled by manipulating the steam input to the heater. The temperature of the liquid feed to the heater is measured and is available to the control system as a disturbance input. Thus, through the use of feedforward control it is possible to correct for changes in the feed temperature before they impact the outlet temperature. See the Appendix for directions on using the web-based interface.

Step 1: Using the web-based interface, open the Feedforward Control workshop. In the workspace shown in Figure 13-10, make a change in the process disturbance value with the PID in Automatic (with Feedforward disabled). What impact does the change in the disturbance value have on the process and the PID output value needed to maintain setpoint?

Step 2: Change the mode of the PID to Manual. Identify the process gain, lag, and deadtime for a step change in the PID output.

Step 3: With the PID in Manual and the process at steady state, change the value of the process disturbance input by a step and identify the process gain, lag, and deadtime for a change in the disturbance input.

Step 4: Calculate the feedforward gain (see lead/lag setup for guidance) based on the process responses. Enter this gain in the FF_GAIN parameter of the PID and enable feedforward action through the FF_ENABLE parameter. Note: the parameter to enable PID feedforward can only be changed when the PID is in manual or out-of-service mode.

Step 5: Change the mode of the PID to Auto and let the loop settle at the SP value. Make a change in the disturbance input and observe the controller and process output. Did the feedforward action minimize the impact of the disturbance?

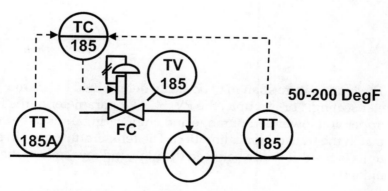

Figure 13-9. Process for Feedforward Control Workshop

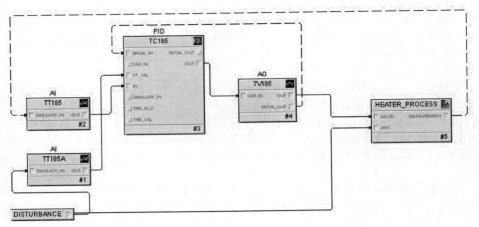

Figure 13-10. Feedforward Control Workshop Module

13.1.4 Workshop Discussion

The feedforward workshop allows you to explore the implementation and use of feedforward control. It is also designed to reinforce previously covered ideas associated with process characterization. For example, it is important to remember that when calculating process gain, the changes in process input and output should be expressed in the same units, for example, percent of range or same engineering units such as °F. The manipulated parameter in this example has process units of % of range. However, as is typical in a plant installation, the measurements associated with the outlet and inlet temperature are in engineering units °F. Thus, to calculate the process gain associated with the manipulated input, it is necessary to convert the change in outlet temperature to percent change in temperature as shown below.

$$\Delta Temperature(\%) = \Delta Temperature(EU) \times \left(\frac{100\%}{Temperature\ EU\ span}\right)$$

The temperature span may be easily determined by examining the parameter scaling. For example, the PV_SCALE parameter of the PID allows the upper and lower end of range and engineering units to be configured to match the transmitter calibration. The temperature span in engineering units is the difference between the high and low ends of the engineering units range.

Temperature EU span = Upper end of range (EU) – lower end of range (EU)

Thus, the process gain associated with the manipulated parameter, the steam input flow rate, can be calculated by putting the PID in Manual and making a step change in the PID output to change the implied valve position. By noting the change in PID output and the resulting change in outlet temperature, the process gain associated with the manipulated parameter may be calculated as shown below.

$$G_m = \left(\frac{\Delta Outlet\ Temperature(\%)}{\Delta PID\ Output(\%)} \right)$$

The disturbance input, inlet heater temperature, is measured in engineering units, ℉ . Thus, since the controlled parameter, outlet temperature, is in the same units, the process gain associated with a change in the disturbance input can be calculated without converting the measurements as shown below.

$$G_d = \left(\frac{\Delta Outlet\ Temperature(°F)}{\Delta Inlet\ Temperature(°F)} \right)$$

The process gains for a change in the manipulated parameter and for a change in the disturbance input should have been determined to be the following:

$$G_m = 1.0$$

$$G_d = 0.3$$

The gains calculated based on the observed step responses may differ slightly from the values shown above. As noted above, G_m is positive because when steam flow to the heater is increased (with feed flow rate held constant), the outlet temperature will increase. Similarly, G_d is positive since for a constant steam input an increase in the inlet temperature will cause the outlet temperature to increase. Thus, the feedforward gain is:

$$FF_{gain} = -\left(\frac{G_d}{G_m} \right) = -0.3$$

Dynamic compensation of the feedforward control was not required in this example since the deadtime and time constants were the same for a change in the manipulated parameter and the load disturbance.

When the load disturbance, heater inlet temperature, was increased with feedforward control disabled, then the PID took no action until the controlled parameter deviated from setpoint. To bring the temperature back to setpoint the PID had to cut back on steam input to compensate for this increase in inlet temperature. However, with feedforward enabled the same inlet temperature change had no impact on the outlet temperature, that is, the steam valve position was changed by feedforward action to compensate for the change in inlet temperature.

As illustrated by this example, feedforward can be an effective tool in reducing the impact of process disturbances. When a process frequently experiences large process disturbances and variations in the process control provided using a PID must be minimized, the use of feedforward may be justified. As mentioned earlier in this chapter, the main costs of adding feedforward control are those associated with purchasing, installing, and maintaining a measurement device for the load disturbance (if not currently measured) and the time required to engineer and commission the feedforward control.

During commissioning of a feedforward system, the disturbance may be something that can be changed to examine the process response. For example, the disturbance may be the throughput of the process. In this case, the operator of the process unit involved may increase the throughput by a step to allow the response to be characterized. In the workshop, the disturbance input could be changed by a step. This allows the deadtime and time constant associated with the disturbance to be quickly determined.

In an operating plant, the major process disturbances may not be measured. The instrument department may be reluctant to install a transmitter to measure a disturbance because of the cost of purchasing, installing, and maintaining a new transmitter. So, the cost justification for installing feedforward control must be based on the benefits of reduced variation in the controlled parameter through the addition of feedforward control.

13.2 Cascade Control

Cascade control may be applied when a process is composed of two or more (sub)processes in series. To apply cascade control, the output of each of these processes must be measured. The input to the first process is directly manipulated. Since the output of each process in the series is the primary input of the next process in the series, any change in the manipulated input to the first process in the series will impact the output of the other processes. The output of each process in the series is the controlled parameter of the PID associated with that process. The PID blocks that make up a cascade control strategy are commonly referred to in the following manner:

Cascade inner, secondary, or slave loop – PID in a cascade strategy that manipulates the process input.

Cascade outer, primary, or master loop – PID associated with the final output of the processes making up a cascade.

Cascade intermediate loops – PIDs positioned between the outer and inner loops of a cascade.

Primary and intermediate loops are configured to adjust the setpoint of the PID associated with the upstream process in the series. An example of cascade control for a process made up of two (sub)processes in series, which is known as a dual loop cascade, is illustrated in Figure 13-11.

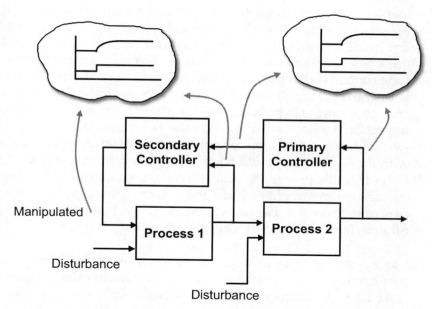

Figure 13-11. Basis of Cascade Control

The PIDs that make up a cascade loop must be commissioned in the order of the process units, starting with the slave loop. In automatic and manual mode, the slave PID of a cascade behaves like a single-loop controller. The tuning techniques previously discussed for single loop may be used to tune the slave PID. Once the slave loop has been tuned for fast response, then the mode of the slave PID should be set to Cascade to allow the PID for the next process in the series to be tuned.

In this dual loop cascade example, the master PID output determines the setpoint of the slave PID. The response of the control parameter of the master PID for a step change in the master PID output reflects the response of processes 1 and 2 as well as the tuning of the slave controller. It is critical that the slave PID controller be tuned before any attempts are made to tune the master PID controller. The slave PID is considered to be part of the process equipment addressed by the master controller. Therefore, after the slave controller is tuned and placed in cascade control, the master controller may be tuned using the tuning techniques previously discussed for single-loop control. In this case, the manipulated parameter is the setpoint of the slave loop and the controlled parameter is the output of Process 2.

13.2.1 Benefits

The process example used to illustrate the concepts of cascade control could have been implemented using single-loop control. If it were implemented as a single loop, the manipulated parameter would be the input to Process 1 and the controlled parameter would be the output of Process 2. The question might be, "Why go to this level of complexity? Why not just use single-loop control?" One of the main reasons for implementing cascade control is that the PID at each point in the cascade can react quickly to disturbance inputs to its associated process. If the PID responds quickly enough, changes introduced by disturbance inputs will have little or no impact on the downstream processes. For example, the rule of thumb when applying dual loop cascade control is that the process associated with the slave loop should have a response time that is at least four times faster than the process associated with the master loop. In this example, cascade control might be justifiable if there were significant load disturbances to Process 1. However, cascade control provides no advantage in minimizing the impact of load disturbances to Process 2.

Cascade control may be implemented in some cases to compensate for the non-linear installed characteristic of a regulating valve. For example, the slave loop of a cascade might be associated with a flow process where the installed characteristic of the valve is non-linear. Since the flow process is capable of very fast changes in flow rate, the slave loop could be tuned to quickly adjust the valve to achieve the flow rate setpoint requested by the master loop. The non-linear installed characteristic of the valve would have no impact on the tuning or response of the master loop. This is a real benefit if the process associated with the master loop is very slow to change. For example, control performance of a slow responding temperature process could be improved by the application of cascade control since a non-linear installed characteristic and any disturbances to the flow process have no impact on the master loop.

As with feedforward control, the added cost of implementing cascade control is associated with the purchase, installation, and maintenance of one or more transmitters to measure process outputs that may not be available or required for single-loop control. More time is required to engineer and commission a cascade control strategy. Also, time may be required to train operators on how to interact with a cascade control strategy. The process improvements that may be achieved using cascade control must justify this added expense.

13.2.2 Example – Superheater Temperature Control

Utility boilers located in the powerhouse area of a plant are often used to generate the steam needed to operate a process. The temperature of the steam supplied by these boilers can have a large impact on process operation. By using an attemperator, in which steam is mixed with water, the temperature of steam exiting the boiler may be regulated. A valve, commonly referred to as the spray valve, is used to adjust the flow rate of water introduced into the attemperator. This steam temperature control application is an example that demonstrates how cascade control may be used to minimize the impact of changes in steam exiting the boiler and other disturbances.

Inside the attemperator, the water mixes with the inlet steam flow. The steam within the attemperator is cooled as the water is converted into steam. The cooled steam from the attemperator flows through a tube bank known as the superheater section of the boiler. In the superheater, the steam is heated by the hot gases passing around the superheater tubes. The steam leaving the boiler is thus superheated.

> **Superheated Steam** - Steam whose temperature, at any given pressure, is higher than water's boiling point.

This cooling and heating process is illustrated in Figure 13-12.

Any change in the spray valve position is very quickly reflected in the temperature of the steam exiting the attemperator. However, a change in the attemperator outlet temperature is not immediately reflected in the boiler outlet steam temperature because the steam must travel through the superheater tube section before leaving the boiler. This long transport delay makes it very difficult to control the superheater outlet steam temperature using single-loop control. For example, changes in the steam flow rate and energy input to the boiler can have a significant impact on the attemperator outlet temperature. With single-loop control, no change in the spray valve position will be made until this disturbance impacts the steam temperature measured at the boiler outlet.

When cascade control is used for boiler outlet temperature control, any change in the steam temperature leaving the attemperator is acted on

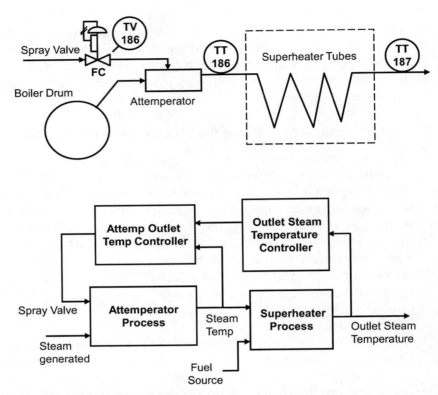

Figure 13-12. Example – Control of Outlet Steam Temperature

immediately by the slave loop of the cascade and thus has no impact on the boiler outlet temperature.

The slave loop of the cascade strategy for boiler outlet temperature control manipulates the water spray flow rate based on the temperature of the steam exiting the attemperator. The set point of the slave loop is manipulated by the master loop of the cascade to maintain the boiler outlet steam at a target temperature.

The setpoint of the master loop would be set to a value required for best plant operation. Disturbances that impact the attemperator outlet temperature are quickly acted on by the slave loop to maintain the temperature at that requested by the master loop. The process response to changes in the slave loop setpoint are linear even though the installed characteristic of the spray valve is not linear. The master loop must only contend with disturbances such as boiler gas temperature that impact the superheater process. Thus, the boiler outlet temperature may be tightly controlled by the master loop of the cascade.

13.2.3 Implementation

As we have seen, cascade control may be implemented when a process is made up of a series of processes. The outlet of each process in this series must be available as a process measurement. Also, one PID block is required for each process in the series. The implementation of a dual cascade control is illustrated in Figure 13-13 where the PID, analog input, and analog output blocks are based on Fieldbus Foundation blocks. For normal operations, the PID block used in the master loop is maintained in Automatic mode. The PID used for the slave loop is operated in cascade mode. The PID output of the master loop is connected to the cascade input of the PID of the slave loop. The PID output of the slave loop regulates the process input through an analog output block.

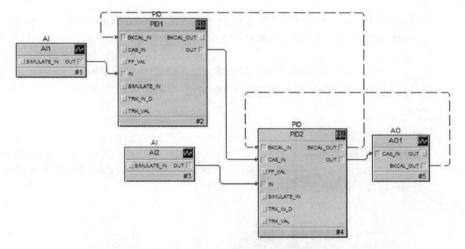

Figure 13-13. Cascade Control Implementation

The convention for function block implementation is that information in the control path flows from left to right. Thus, the PID associated with the master loop appears on the far left and the slave loop is shown on the far right in this drawing. On a Piping and Instrumentation Diagram (P&ID), the process flow may be from left to right. Thus, the order of the process units and the cascade elements in these drawings is often reversed.

The status of the back calculation connections used in a cascade strategy is used to automatically provide bumpless transfer when a PID is changed from manual or Automatic mode to Cascade mode. The back calculation connections between the PID blocks and analog output block are critical for correct operation.

During normal plant operations, it is possible for the analog output block or for the output of the PID blocks in the cascade strategy to become limited. For example, the block setpoint or output has reached a high or low limit value. Through the back calculation connection, the status that is communicated indicates this high and low limit condition. When the back calculation input status indicates that a high or low limit has been hit downstream, the PID integral calculation is modified to prevent windup if the calculated change in output would drive the process further toward the limit. If the change would drive the process away from the limit, then normal integral action is enabled. In the status communicated in the back calculation output, the PID is designed to take into account whether the PID is direct or reverse acting. Thus, use of the back calculation connection by the PID and analog output avoids the need for the user to build special logic to handle limit conditions that may occur in the cascade.

The PID block supplied in many products is designed to support dynamic reset limiting, also commonly know as external reset feedback.

> **External Reset Feedback** – Calculation of the PID integral contribution based on a measurement of the manipulated process input.

When the PID supports this feature, the performance of cascade control loop may be improved by enabling this option. One example of how this option may be enabled in some control systems and field devices is illustrated in Figure 13-14.

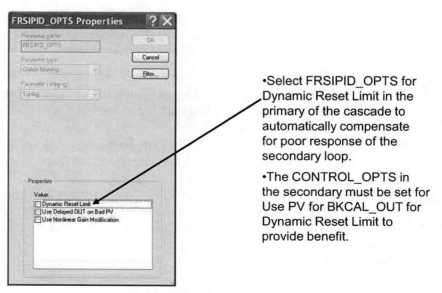

•Select FRSIPID_OPTS for Dynamic Reset Limit in the primary of the cascade to automatically compensate for poor response of the secondary loop.

•The CONTROL_OPTS in the secondary must be set for Use PV for BKCAL_OUT for Dynamic Reset Limit to provide benefit.

Figure 13-14. Dynamic Reset Limiting Option

This option should be selected in all PIDs that make up the cascade control strategy. Also, it is necessary to select the control option Use PV for Back Calculation. When these selections are made, the back calculation input value and status are used in the reset calculation. This added information is used to provide better dynamic response. For example, if the slave loop is not maintaining setpoint because of conservative tuning, process non-linearity, or rate limiting in the analog output, then the integral action of the master loop is automatically adjusted to match the slower response of the slave loop. The control performance of the master loop may thus be improved.

The correct operation of a cascade control strategy depends on the process measurements being available at all times. However, a transmitter may fail. Also, under startup conditions, the flow rates may be so low that flow measurements based on a differential pressure measurement may not be reliable. To avoid having to revert back to manual control under these conditions, the PID Bypass option on an intermediate or slave loop may be enabled. When PID Bypass is enabled, the setpoint of that PID is transferred directly to its output, taking into account direct and reverse and scaling, without any control action from the PID. By enabling PID Bypass in the slave loop of a dual cascade, the control strategy essentially reverts to single-loop control.

13.2.4   Workshop – Cascade Control

This Cascade Control workshop is designed to give you experience operating a cascade control system. Also, the benefits of dynamic reset and PID Bypass are demonstrated through the workshop exercises. See the Appendix for directions on using the web-based interface.

Step 1: Using the web-based interface, open the Cascade Control workshop. In the workspace shown in Figure 13-15, set the mode of the slave loop (PID2) to manual. Make a step change in the OUT of PID2 and observe the response of the two process outputs. What difference is visible in the response of the two outputs?

Step 2: Place PID2 in automatic mode and change the setpoint to 50. Observe the process response and also the automatic tracking that is done by the master loop.

Step 3: Place PID2 in cascade mode. With PID1 in manual mode, change the output of PID1 and observe the change in the setpoint of PID2.

Step 4: Place PID1 in automatic mode and observe the response when a change is made in the setpoint. With PID2 in cascade mode and PID1 in automatic mode, make a step change in the load disturbance. What impact was there on the slave loop? Did the master loop change as a result of the change in load disturbance?

Step 5: Enable FRSI_OPTS, Dynamic Reset Limit in PID1. Enable CONTROL_OPTS Use PV for BKCAL_OUT in PID2 and make a change in the PID1 setpoint. Is there any difference in the response?

Step 6: Enable BYPASS in PID2 and observe the difference in response when the setpoint of PID1 is changed.

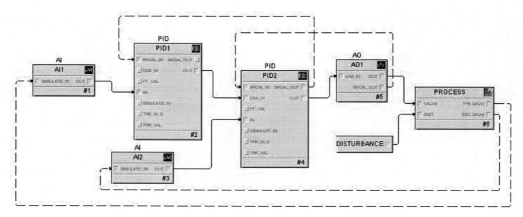

Figure 13-15. Cascade Control Workshop Module

13.3 Override Control

To ensure proper process operating conditions, it may be necessary to maintain some measurable parameters such as vessel temperature, pressure, and level within certain operating limits (constraint limits). The implementation of override control is often the most effective way to maintain the process within its operating constraint limits. This control technique is quite important within the process industry. Tools are provided in most control systems that allow override control to be implemented. Unfortunately, in the past, this technique has often been underutilized.

Override control may be applied to a process that is characterized by having one manipulated parameter, one controlled parameter, and one or more constraint outputs. The implementation of an override control involving one constraint output is illustrated in Figure 13-16.

In general, override control may be implemented using two or more PID blocks and a control selector block. The control selector block defined by Fieldbus Foundation can support as many as three PID blocks. When

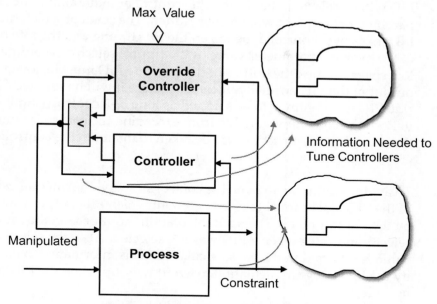

Figure 13-16. Basis – Override Control

more than two constraints must be addressed, then multiple selector blocks may be used in override control.

A common use of override control consists of one controlled parameter and one constraint variable as illustrated in Figure 13-16. Under normal operating conditions, the controlled parameter is maintained at setpoint by the selected PID. The override PID takes an active role if the value of the constraint variable approaches its setpoint. The setpoint of the override PID is set to the constraint limit, that is, a value that the constraint variable should never exceed. The two PID outputs are connected to the selector block input. Based on the values of the PID outputs, one output is selected and is passed to the analog output block. In this example, the lower PID output value is selected. Which PID was selected is reflected in the status of the back calculation connections between the control selector and the PID blocks involved in the override strategy. The control selector may be configured to select the higher or the lower output value.

13.3.1 Override Operation

During normal process operating conditions, the controlled parameter is maintained at setpoint and the constraint variable is far from the constraint limit. The output of the PID associated with the controlled parameter is selected and passed to the analog output block. However, as processing conditions change, the constraint parameter value may start to approach the constraint limit and at some point the output of the override

PID will be selected, that is, the PID controller maintaining the controlled parameter at setpoint will be over-ridden. The point at which the override PID takes over control depends on the PID tuning and the rate of change of the override parameter value. If the control selector is configured for low selection then the output of the override PID must be lower than the output of the PID normally in control before it will be selected. The override PID will continue to be selected as long as a control action is needed to keep the constraint variable from exceeding the constraint limit. Under these conditions, no action will be taken to maintain the controlled parameter at setpoint.

To avoid windup in an override control strategy, the integral calculation of the PID blocks must take into account which PID output was selected by the selector block. The back calculation input provided by the control selector indicates when a PID was not selected. Also, the selected output value is reflected in the back calculation. This information is used to modify the PID integral calculation to avoid windup when the PID output is not selected.

If the PID blocks support dynamic reset, the best process response may be achieved by enabling dynamic reset in the PID blocks used in an override control strategy. When dynamic reset is enabled, then the value of the selector output provided by the back calculation input is used in the integral calculation.

To establish the tuning of the PID associated with the controlled parameter, the mode of the override PID may be changed to Manual and its output changed to a value that will not be selected. At this point, the PID for normal control can be tuned using the same techniques previously described for single-loop control. Once the PID for the controlled parameter has been tuned, the output of the override PID can be manually changed to match the selected output. The PID for the controlled parameter can be placed in Manual and its output changed to a value that will not be selected. The override PID output should now be selected and the tuning can be established.

Once all the PIDs in the override strategy have been tuned, their setpoints may be changed to their normal value and the PIDs placed in Automatic. The order in which this is done should be chosen to avoid any sudden changes in the selected output.

13.3.2   Example – White Liquor Clarifier

Clarifiers are used in the process industry to separate a liquid from entrained solid components. A white liquor clarifier is used to separate a caustic liquid, sodium hydroxide (white liquor), from suspended particles of calcium carbonate. Inside the white liquor clarifier, calcium carbonate settles to the bottom of the clarifier and is transferred to a storage tank using a variable speed pump. When the transfer flow rate exceeds the clarifier settling rate then at some point all sediment will have been removed and no further pumping should take place. Since there is a difference in the density of calcium carbonate and white liquor, this condition may be detected by measuring the density of the material in the transfer line. Under normal conditions, a variable speed pump is adjusted to maintain a target flow rate setpoint. However, if the density decreases, indicating a low concentration of calcium carbonate, then the variable pump speed must be reduced to prevent the concentration from dropping below the minimum concentration that may be pumped to storage. These control objectives may be achieved using override control as illustrated in Figure 13-17.

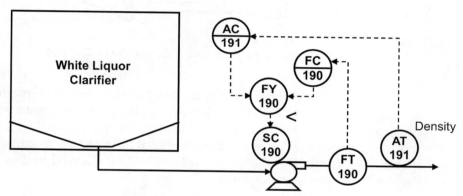

Figure 13-17. Example – Clarifier Control

In this example, the constraint parameter is the transfer line material density. The setpoint of the override PID is the minimum material density that can be transferred to storage. If a drop in density measurement exceeds this constraint limit, then the density controller will override the flow controller and reduce the variable speed pump to honor the constraint limit.

13.3.3 Example – Compressor

In this example of override control, a large natural gas compressor is powered by an electric motor. Under normal operating conditions, the gas flow to the compressor is regulated to maintain a constant discharge pressure. As the demand for gas increases, the inlet flow rate must be increased to maintain discharge pressure. The load on the electric motor changes with the gas flow rate to the compressor. As the gas flow and the associated load on the compressor increase, then the electric current to the motor also increases. If the current exceeds some limit, then this could damage the motor. In this example, the motor current is the constraint variable and the discharge pressure is the controlled parameter. The compressor override strategy is shown in Figure 13-18.

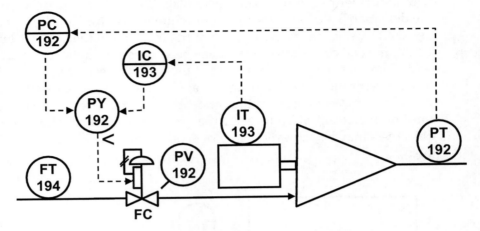

Figure 13-18. Example – Compressor Control

For typical gas flow demands, the discharge pressure can be maintained by adjusting the inlet flow rate. However, if the motor current exceeds the constraint limit set by the current controller, then the inlet flow will be reduced to keep the current at the constraint limits to avoid damaging the motor. As a result, the discharge pressure will be allowed to drop below the discharge pressure setpoint.

13.3.4 Implementation

An override strategy involving one constraint variable can be implemented using a control selector block, two PID blocks and associated analog input and output blocks as illustrated in Figure 13-19.

The control selector block defined by Foundation Fieldbus supports upstream and downstream back calculation connections. Numbered pairs of input and back calculation outputs of the control selector should be connected to the same PID. Dynamic reset is supported by the PID and should be enabled. The control selector may be configured as a high or low selection. When more than three PIDs are needed in an override strategy, then multiple control selector blocks may be used. For example, two control selectors could be used to address four inputs where the output of one selector becomes an input to the second selector. As illustrated in Figure 13-20, the control selector back calculation status is based on the mode of the control selector block and which output was selected.

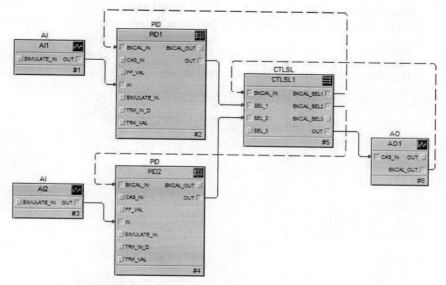

Figure 13-19. Typical Override Control Implementation

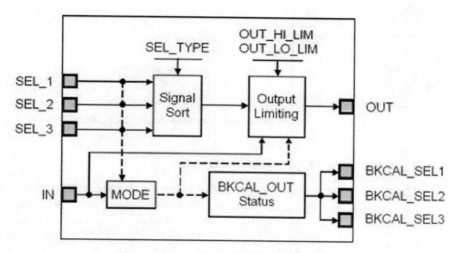

Figure 13-20. Control Selector Block

13.3.5 Workshop – Override Control

This Override Control workshop is designed to allow the operation of override control to be explored. It is assumed that the PIDs have been tuned. See the Appendix for directions on using the web-based interface.

Step 1: Using the web-based interface, open the Override Control workshop. In the workspace shown in Figure 13-21, place the two PIDs in Auto and observe the response when both setpoints are set to 50.

Step 2: Change the setpoint of PID1 to 60 and observe the response of PID2. Change the setpoint of PID1 to 45 and observe the response.

Step 3: Change the disturbance input to 30 and observe the response of the override loop and the process response.

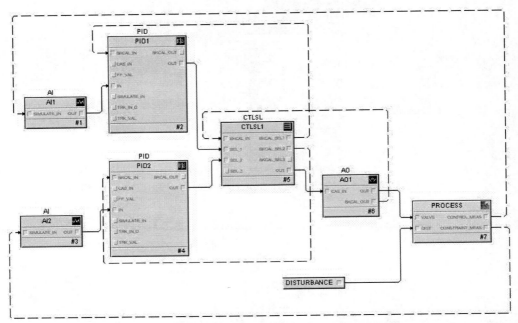

Figure 13-21 Override Control Workshop Module

13.3.6 Workshop Discussion

The main purpose of this workshop was to demonstrate override control. The control selector was set up with a high selection. The choice of high or low selection depends upon the control objective. When working with an override strategy, there are a number of indications of which PID output has been selected. If the controlled parameter is at setpoint, then this would indicate the associated PID has been selected. We can confirm this by verifying that the PID output value matches the control selector output.

As demonstrated in this workshop, the output that is selected in an override strategy may change with a PID setpoint. Also, process disturbances can impact which PID is selected.

13.4 Control Using Two Manipulated Parameters

A wide variety of processes may be described as having one controlled parameter and two or more manipulated process inputs as illustrated in Figure 13-22. The challenge from a control perspective is that there is no unique set of input values that are required to maintain the controlled parameter at setpoint. For the example of two manipulated inputs, the controlled parameter may be maintained at a particular setpoint value when both inputs are at 50%. However, the same controlled parameter value may be achieved with other input combinations such as 45% and 60% or 30% and 80%. For all practical purposes, there are an infinite number of input combinations that may be used to achieve a setpoint. In such cases, the requirements for control implementation may be described as being "underspecified" or having extra degrees of freedom.

> **Underspecified** – To give insufficient, or insufficiently precise, information to specify completely.

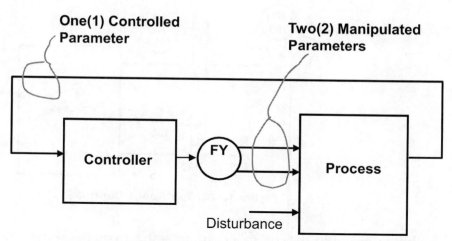

Figure 13-22. Control Using Two Manipulated Parameters

The control implementation requirements may be defined through the use of a control element (FY) which maps the controller output to the manipulated process inputs in a unique manner. The most common techniques for implementing control using two manipulated process inputs are addressed in the remaining portion of this chapter:

Split-range control
Valve position control
Ratio control

The implementation and control response for setpoint and changes in disturbance inputs provided by each technique will be examined in some detail. Also, examples are used to illustrate where such techniques have been successfully applied in the process industry.

13.4.1   Split-range Control

One of the most common ways of addressing multiple process inputs is known as split-range control. Using this approach, a splitter block is used to map the controller output to multiple manipulated process inputs. The splitter block may be used to define a fixed relationship between the controller output and each manipulated process input as illustrated in Figure 13-23.

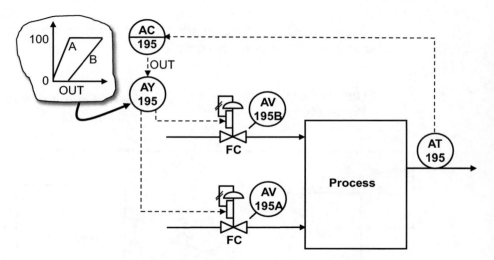

Figure 13-23. Split-range Control

In this example, the controller output is shown on the *x*-axis. The input values to the two valves as calculated by the splitter are labeled in this figure as A and B. The splitter defines how each valve is sequenced as the controller output changes from 0 to 100%. From the controller's perspective, it appears as though there is only one manipulated process input and the splitter is considered to be part of the process. Once the splitter has been configured, the PID block used with the splitter can be commissioned and operated in the same manner as a single-loop controller. Through the use of the splitter block, the two valves are sequenced in a fixed way so that they essentially look like one valve to the control. The various ways in which the valves may be sequenced using the splitter block will be illustrated through a number of examples.

Power House Example

Steam turbines used to generate electricity are powered by high pressure
steam generated by one or more boilers located in the powerhouse area of
the plant. The boiler steam generation demand is regulated by the plant
master controller to maintain a constant pressure in the main steam
header used to supply the turbine steam. For example, in newer installations, the supply header may be maintained at 1475 psi.

The pressure of the steam is reduced as it flows through the turbine,
allowing lower pressure steam to be extracted at various points. This
lower pressure steam may be used to meet the plant's process steam
demands. For example, at one point, the extraction may be regulated to
meet the steam demands of a 400 psi header. Extraction valves associated
with the turbine are automatically regulated to maintain the lower header
pressure constant.

To allow the plant to continue operation if the turbine or generator fails
and must be shut down, pressure reducing valves (PRVs) between the
high pressure header and the lower pressure header may be adjusted to
meet the lower pressure header steam demand and to maintain the header
pressure constant. Multiple valves may be required to allow the steam
flow to be precisely adjusted over a wide operating range. This may be
accomplished by using a splitter block in conjunction with a PID block to
adjust the pressure reducing valves. An example of how this has been
implemented with two valves is illustrated in Figure 13-24.

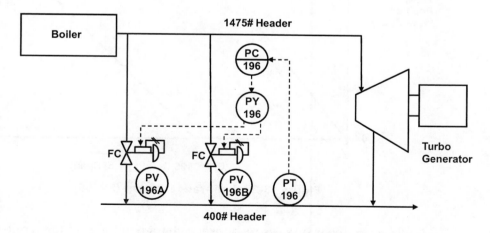

Figure 13-24. Steam Header Example

When a fault is detected in the generator or turbine, the steam flow to the
turbine is automatically shut off by blocking valves. The high pressure
header then starts to build up in pressure and the controls used to main-

tain the high pressure immediately reduce the firing rate of the boiler. At the same time, the pressure in the 400-psi steam header begins to drop from the loss of steam from turbine extraction. To allow both disturbances to be automatically corrected, the setpoint of the lower header pressure controller is set just below the header pressure maintained when the turbine is on-line. When the pressure drops below this setpoint, indicating that the turbine is not on-line, the lower header pressure controller output changes to open the pressure reducing valves. As a consequence, the flow of high pressure steam is re-established and the lower header pressure is maintained at a constant value.

From the perspective of the PID block used to maintain the lower header pressure, the splitter block should make the two valves respond as one valve that provides the resolution of the smaller valve. Thus, the splitter block should be configured to open one valve until it is fully open, and then open the other valve. For this example, valve 196A is opened first, and then, as the controller output increases, valve 196B is opened as illustrated in Figure 13-25.

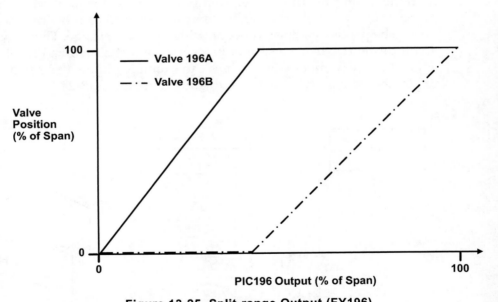

Figure 13-25. Split-range Output (FY196)

To achieve a linear process response, a change in the pressure controller output at any point over its operating range should result in the same change in steam flow to the low pressure header. For example, a 1% change with a controller output of 30% should produce the same change in steam flow as a 1% change with a controller output of 80%. When the splitter sequences the valves in this manner, the process gain will be constant.

Splitter Characterization

For split-range sequencing, the range of the controller output over which each valve opens and closes must be defined as part of the splitter config-uration. If the valves adjusted by the splitter block are exposed to the same operating conditions (and are, of course, identical in design), then it is only necessary to know the sizes of the valves to configure the splitter block to provide a linear response. In the steam header example, the inlet and outlet pressure are the same for the two pressure reducing valves. The maximum steam flow in pounds per hour (lb/hr) through either valve is determined by the valve Cv at 100% open, steam temperature and quality, and the steam pressure in the 1475 psi header and the 400 psi header.

When the valves used in split-range control are the same size, then the splitter block output ranges over which each valve is adjusted would be defined equal. For example, as the output changes 0–50%, adjust one valve from 0–100% and for output changes of 50–100%, adjust the second valve from 0–100% while maintaining the first valve open. However, when the valve sizes or operating conditions for the valves are different, then it is necessary to define these ranges to compensate for these differences. By correctly defining the controller output regions over which each valve operates, then at any operating point, a change in the output should pro-duce the same response in the controlled parameter.

When the steam flow rating for full open position at the expected operat-ing conditions of each valve is known, the operating range of the control-ler output for the first valve in a split-range control is calculated as follows:

$$Output\ Range_{valve\ 1} = 100 \times \left(\frac{Valve\ 1\ flow\ rating}{Valve\ 1\ flow\ rating + Valve\ 2\ flow\ rating} \right)$$

For example, if the flow ratings in thousand of pounds per hour, KPPH, of the valves used in split-range control were as shown below:

$$Valve\ 1\ flow\ rating = 50\ KPPH$$

$$Valve\ 2\ flow\ rating = 150\ KPPH$$

Then the controller output range of adjustment associated with Valve 1 would be:

$$Output \ Range_{valve \ 1} \ = \ 100 \times \left(\frac{50}{50 + 150}\right) = 25\%$$

If Valve 1 is opened first, then the splitter block would be configured in the following manner:

Controller Output (%)	Valve 1 (%)	Valve 2 (%)
0–25	0–100	0
25–100	100	0–100

When the controller output is operating in the 0–25% range, then a 1% change in the controller output would result in the following change in Valve 1:

$$Valve1 \ \%change \ = \ 1\% \ Controller \ Output \times \left(\frac{100}{25}\right) = 4\%$$

$$Valve1 \ \%change \ = \ 1\% \ Controller \ Output \times \left(\frac{100}{25}\right) \times \left(\frac{50KPH}{100\%}\right) = 2KPPH$$

A 1% change in the controller output when it is operating in the 25–100% range will produce the following change in Valve 2:

$$Valve2 \ \%change \ = \ 1\% \ Controller \ Output \times \left(\frac{100}{75}\right) = 1.33\%$$

$$Valve2 \ \%change \ = \ 1\% \ Controller \ Output \times \left(\frac{100}{75}\right) \times \left(\frac{150KPH}{100\%}\right) = 2KPPH$$

By configuring the splitter in this manner, a 1% change in the controller output results in the same change in flow rate over the entire operating range of the controller output.

Example - Slaker

The temperature control of green liquor flow to the slaking operation in the pulp and paper industry is another example of split-range control. In this application, the temperature of green liquor added to the slaker is maintained at setpoint by a temperature controller. The cooling water and steam provided to a cooler and heater, respectively, are adjusted in a split-range manner to achieve the setpoint temperature, as illustrated in Figure 13-26.

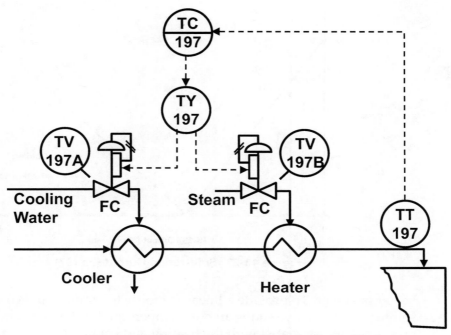

Figure 13-26. Example – Split-range Control

Changes in the green liquor feed temperature and flow rate are the primary disturbance to this temperature. Also, changes in the steam or cooling water temperature are unmeasured process disturbances. If the inlet green liquor temperature falls below the temperature controller setpoint, then the steam flow to the heater is increased. If the feed temperature rises above the temperature controller setpoint, then the water flow to the cooler is increased.

The heater and cooler are never used at the same time. Thus, when steam is added to the heater, the cooling water valve must be closed. Similarly, when the cooling water valve is open the steam valve must be closed. The heater and cooler valves may be sequenced in this manner using the splitter characterization illustrated in Figure 13-27.

The cooling valve is adjusted at the lower range of the temperature controller output. During this time, the heating valve is closed. In the upper range of the temperature controller output the heating valve is adjusted and the cooling valve is closed. At the upper end of the cooling range and the lower end of the heating range, both the heating and cooling valves are closed.

The same change in the position of the heating and cooling valves may result in a different amount of heat energy being added or removed by the

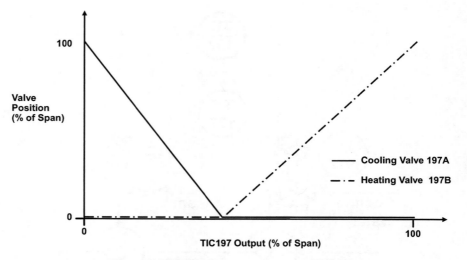

Figure 13-27. Split-range Output (FY197)

heater or cooler. To provide a linear response to temperature control, the splitter block must be configured to compensate for the difference in the process gains associated with the heating and cooling valves. The process gains may be quickly determined by observing how a step change in a valve position impacts the temperature of green liquor flow to the slaker. For example, the heating valve position may be changed 1% while the cooling valve is maintained closed. By measuring the resulting change in temperature, the process gain associated with the heating valve can be calculated. Similarly, the process gain associated with the cooling valve can be calculated based on the observed temperature change for a 1% change in the cooling valve position, as illustrated in Figure 13-28.

The temperature controller output range for the cooling valve is calculated in the following manner:

$$Output\ Range_{cooling\ valve} = 100 \times \left(\frac{\Delta Temp_{cooling\ valve}}{\Delta Temp_{cooling\ valve} + \Delta Temp_{heating\ valve}} \right)$$

$$Output\ Range_{cooling\ valve} = 100 \times \left(\frac{2.2}{1.1 + 2.2} \right) = 66\%$$

Example: Temperature controlled using heating and cooling valves

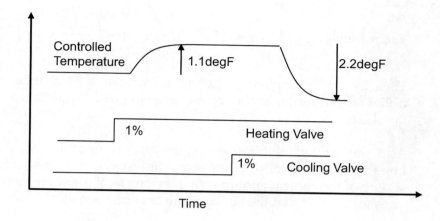

Desired Splitter Span cooling valve = 100*(2.2/(1.1+2.2)) = 66%

Splitter range for cooling valve = 0-66%
Splitter range for heating valve = 66-100%

Figure 13-28. Testing Process to Determine Splitter SP

The splitter block could be characterized in the following manner:

Controller Output	Valve 1	Valve 2
0–66	100–0	0
66–100	0	0–100

When the controller output is operating in the 0–66% range, then a 1% change in the controller output would result in the following changes in the cooling valve position and green liquor temperature:

$$Valve\ Change_{cooling\ valve} = 1\%\ Cont\ Output \times \left(\frac{100}{66}\right) = 1.5\%$$

$$Temp\ Change_{cooling\ valve} = 1\%\ Cont\ Output \times \left(\frac{100}{66}\right) \times \left(\frac{2.2°}{1\%}\right) = 3.3°$$

A 1% change in the controller output when it is operating in the 66–100% range will produce the following changes in heating valve position and green liquor temperature:

$$Valve\ Change_{heating\ valve} = 1\%\ Cont\ Output \times \left(\frac{100}{100-66}\right) = 3\%$$

$$Temp\ Change_{heating\ valve} = 1\%\ Cont\ Output \times \left(\frac{100}{100-66}\right) \times \left(\frac{1.1°}{1\%}\right) = 3.3°$$

Thus, by configuring the splitter in this manner, a 1% change in the controller output results in the same change in green liquor temperature to the slaker.

As demonstrated by this example, the process input that has the smaller impact on the controlled parameter is adjusted over a smaller range. The scaling of the controller output that is done by the splitter then compensates for the fact that the temperature change is smaller.

Unfortunately, in some cases where split-range control has been implemented, the splitter is configured as 0–50%, 50–100% even though there may be a large difference in how each manipulated input impacts the controlled parameter. As a consequence, the process gain as seen by the controller depends on which input is being adjusted. In other words, a nonlinear response will be observed. This means that tuning that works well in one portion of the controller output operating range may result in sluggish or unstable operation in another range of the output. To provide the same response for each region, it is necessary to configure the splitter block based on the impact each manipulated parameter has on the controlled parameter.

Implementation

When split-range control is implemented, a splitter block is used with the PID block as illustrated in Figure 13-29.

The splitter block is designed to use the back calculation outputs provided by the analog output blocks to provide bumpless transfer when an analog output block mode is changed from Auto to Cascade. Also, the mode and limit status of the analog output block is reflected in the back calculation status and value provided to the PID block. Thus, bumpless transfer is provided when the splitter block mode transitions from Auto to Cascade. Also, the status information is used to avoid PID windup when the splitter output is limited.

If neither analog output block is in Cascade mode, then the PID output will be automatically initialized based on the first analog output block that is switched to Cascade mode. Since the splitter sequences the outputs in a fixed relationship when the second output is placed in Cascade mode,

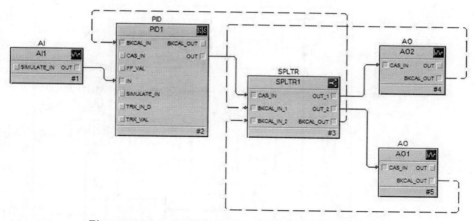

Figure 13-29. Split-range Control Implementation

then the output value may not match the value that would normally be provided by the splitter. To provide bumpless transfer of the second output, the splitter automatically calculates the difference in the current (i.e., now existing) output and the calculated output. This difference is then added as a bias to the calculated output. This bias is then reduced to zero at a constant rate over the configured balance time.

Once split-range control has been commissioned, the analog output and splitter blocks normally remain in Cascade mode. The plant operator can interact with the PID in the same manner as with single-loop control.

Splitter Block

The splitter block may be operated in Automatic or Cascade. In Automatic mode, the setpoint may be adjusted by the operator. The splitter outputs are calculated based on this setpoint value and the characterization specified by the configured input range and output range. When the mode is changed to Cascade, the setpoint of the splitter may be adjusted by the upstream PID block. The internal calculation of the splitter block may be illustrated as shown in Figure 13-30.

The splitter block output is determined by the IN_ARRAY and OUT_ARRAY parameters. Using these parameters, the two output ranges for the controller output and corresponding ranges for splitter outputs 1 and 2 may be specified. Some examples of how these parameters are used to configure the splitter are illustrated in Figure 13-31.

As is shown in the three examples illustrated in Figure 13-31, the controller ranges are often defined with some overlap to ensure that there is no gap in the transition due to the calibration of the final control elements. Another feature of the splitter block is shown in the third example. When

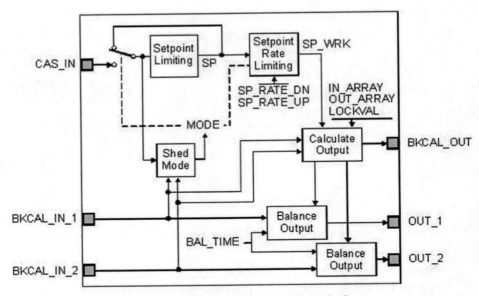

Figure 13-30. Splitter Block Calculation

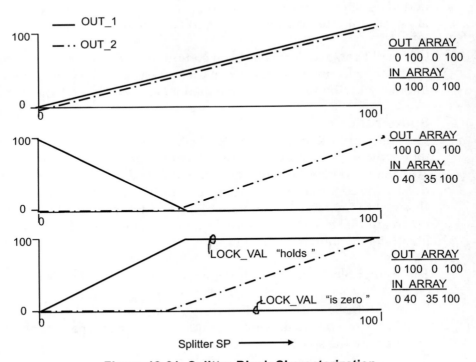

Figure 13-31. Splitter Block Characterization

the controller output transitions from the lower range to the upper range, then the LOCK_VAL parameter may be used to specify that the first output value is either maintained or set to zero. Such a feature may be used when a small valve is used in conjunction with a much larger valve.

An example of a dialog box used to configure the splitter characterization is illustrated in Figure 13-32.

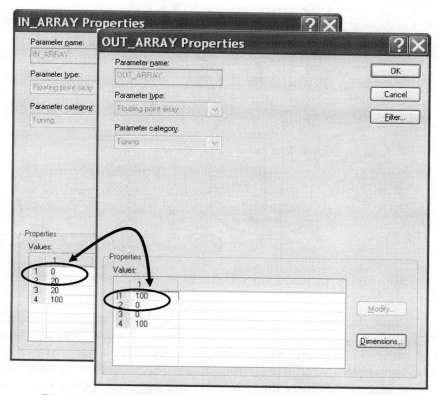

Figure 13-32. Configuring the IN_ARRAY and OUT_ARRAY

As illustrated in Figure 13-32, the first two entries of the arrays define the controller output range and corresponding range of output 1.

Neutralizer Example

The pH variations in a process feed stream or in a discharged liquid by-product of a manufacturing process may be reduced and the pH maintained at a target value using a neutralizer. Because of the strongly non-linear response of pH to the addition of a reagent, it is necessary to be able to precisely adjust the reagent flow to the neutralizer over a wide range. To achieve the resolution needed at low reagent flows, a neutralizer may be instrumented using a small valve for fine (high resolution) adjustment and a larger valve for coarse (low resolution) adjustment. In such a case, the pH may be maintained using split-range control as illustrated in Figure 13-33.

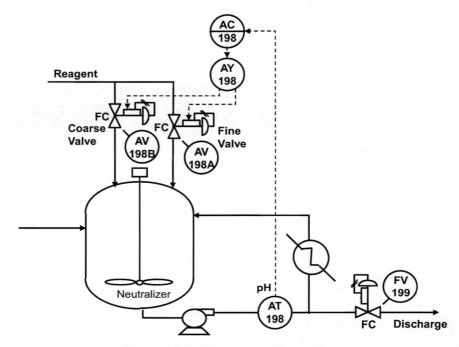

Figure 13-33. Example – Neutralizer

When the demand for reagent is small, the pH is maintained using the small valve. As the reagent demand increases, at some point the small valve will be fully open and the large valve will start to open. To prevent the large valve from operating in a nearly closed position which could potentially damage the valve seat and trim, the splitter may be configured using the LOCK_VAL parameter to close the small valve as the large valve begins to open. In response, the large valve will be opened further—away from the seat—to meet the reagent demand. The configuration of the splitter block for the small and large valve is illustrated in Figure 13-34.

As illustrated in this example, the splitter would be configured to achieve some overlap between the small and large valve. Using the HYSTVAL parameter, some hysteresis may be defined in the points at which the small valve is closed and the point where it is again opened. By specifying the hysteresis value based on the valve opening overlap, it is possible to prevent the small valve from switching on and off around the upper end of its range.

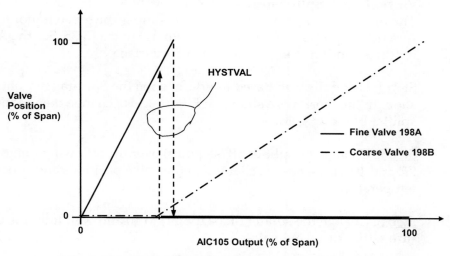

Figure 13-34. Split-range Output – Valve Sequencing

If a control system is not designed to support the functionality provided by the splitter block, then split-range control may be done in the field through the use of an I/P and calibration of the valve actuators. In this case, the pneumatic signal from the I/P will vary over a range of 3 to 15 psi as the controller output changes over its range. The valve actuators used in the control strategy may be calibrated to act upon only part of the 3 to 15 psi signal provided by the I/P. If the valves have an equal impact on the controlled parameter, then the actuator for one valve would be calibrated to operate over a range of 3–9 psi and the actuator for the second valve calibrated to operate over a range of 9–15 psi.

Unfortunately, an engineer responsible for designing the control strategy for a new control system may not be aware that split-range control may be implemented in the control system. If the design is based on a previous installation, then split-range control may be shown as being performed in the field. The primary advantage of implementing split-range in the control system is that the split-range control setup is then much faster to implement and commission. If changes are needed in the split-range sequencing, then this can be done quickly without any changes in field devices. Also, rather than just having the controller output available in the control system, the implied valve position of each value in the split-range control strategy is available for display and diagnostics. The advantage of implementing split-range control in the control system easily justifies the cost of two analog outputs and two wire pairs versus the one analog output and one wire pair needed when split-range is done in the field.

Workshop – Split-range Control

The process example used in this workshop is the green liquor temperature using split-range control as shown in Figure 13-35. See the Appendix for directions on using the web-based interface.

Step 1: Using the web-based interface, open the Split-range Control workshop. In the workspace shown in Figure 13-36, change the mode of the splitter block to Auto.

Step 2: Change the splitter SP (setpoint) over the following range: 0, 25, 50, 75, and 100. Observe the changes in the valve positions and process outlet temperature.

Step 3: Change the splitter SP (setpoint) to 50 and wait until the temperature settles to a fixed value.

Step 4: Make a step change in the FEED_TEMP disturbance and manually adjust the splitter setpoint to get the OUT_TEMP back to its initial value.

Step 5: Change the splitter mode to Cascade, change the temperature control setpoint and observe the response.

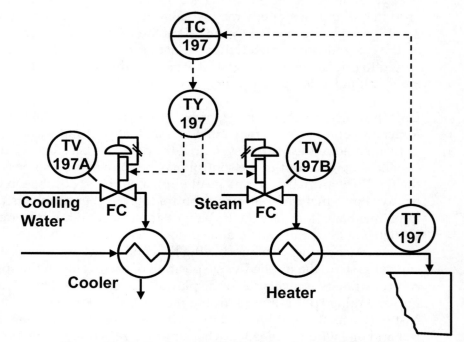

Figure 13-35. Process for Split-range Control Workshop

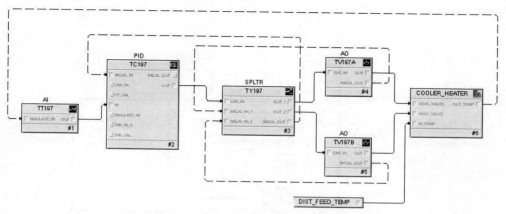

Figure 13-36. Split-range Control Workshop Module

13.4.2 Valve Position Control

The precise adjustment of process feedstock flow is often necessary to maintain the processing conditions required to meet product specifications. To meet the target plant production rates, a high feedstock flow rate is often required. If a single valve is used to regulate the feedstock flow, then the precision with which the flow rate may be regulated is determined by the valve position resolution. The position resolution of a large valve may be the same as that of a small valve in terms of percent of maximum flow. However, as the valve gets bigger, that percent of scale represents a much larger flow rate. Precise metering requirements may be met with a small valve, but a small valve may not have the flow capacity needed to maintain the plant product rate. An ideal solution would be to use a small valve to make fine changes, and use a larger valve to address the need for flow capacity. Split-range control cannot be used to satisfy this requirement since the valves are sequenced one at a time, that is, one valve is used and then the other valve. To achieve precise flow rate control over a large flow range, it is necessary to use both valves at the same time. As fine adjustments in flow are made with the smaller valve, the large valve is automatically adjusted to keep the small valve within its operation range of 0-100%. Valve position control may be used to achieve this type of regulation of the small and large valve.

Valve position control is a straightforward concept that may be easily implemented in most modern control systems. The basic components of a valve position control strategy are illustrated in Figure 13-37.

In this example, the feedback controller directly adjusts the small valve input to the process. The output of the feedback controller—the implied valve position of the smaller valve—is also the control parameter of an integral-only (I-only) controller. In this example, the setpoint of the I-only

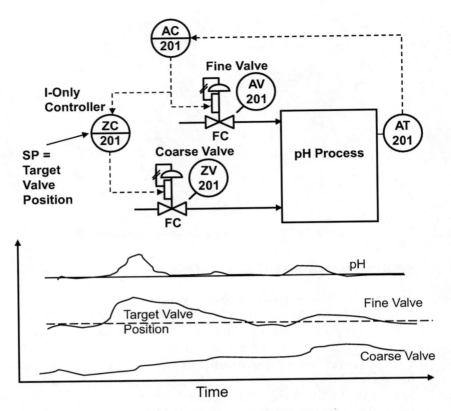

Figure 13-37. Valve Position Control

is set to 50%. If the feedback controller moves the small valve away from 50% open for a sufficiently long period of time, then the I-only controller will gradually adjust the large valve in a manner that drives the small valve back to its normal operating point of 50%. At an operating point of 50%, the feedback controller can make adjustments in either direction to achieve its setpoint. The tuning of the I-only controller is very slow and so a change in the large valve appears as a load disturbance to the feedback controller and is quickly corrected through adjustment of the small valve.

The I-only controller used in the implementation of valve position control is a normal PID but is tuned only using the integral component, that is, the PID is configured for no proportional or derivative contribution. The implied valve position is provided rather than a process measurement as the controlled parameter. As a result, the larger valve's position will be changed slowly, in a very smooth manner. The integral time of the I-only controller should typically be five times longer than the integral time of the feedback controller.

Note: When implementing valve position control, it is extremely important the the I-only controller maintain its last output value if the feedback controller is not in Automatic mode.

Example – Boiler BTU Demand

Valve position control may be used to achieve various control objectives that cannot be met using split-range control. One example is a boiler application, where valve position control is used to minimize the use of costly high BTU fuel by maximizing the use of low cost, low BTU waste fuel. The manner in which this may be achieved using valve position control is illustrated in Figure 13-38.

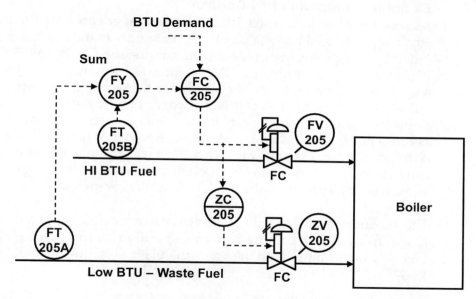

Figure 13-38. Example – Boiler BTU Demand

The boiler in this example is used to generate steam for the plant. The BTU demand from the combustion control system sets the setpoint of the Total BTU controller that adjusts the fuel flow to the boiler. Any change in the high BTU fuel flow rate has an immediate impact on the rate of boiler steam generation, which a change in the flow rate of low BTU waste fuel does not, and thus high BTU is the preferred fuel. However, in this example, high BTU fuel is much more expensive than low BTU waste fuel. To achieve a fast response to changes in BTU demand, the Total BTU controller directly adjusts the high BTU fuel valve. The output of the Total BTU controller is also the input to an I-only controller that adjusts the low BTU waste fuel valve. The setpoint of the I-only controller is set to a value that allows both increases and decreases in BTU demand to be quickly met using high BTU fuel. For example, if the setpoint were set to 30%, then long term the high value BTU valve operate around a value of 30% and thus could response to increase and decrease demands for fuel.

In this example, the flow capacity of the HI BTU valve is much smaller than the flow capacity of the LO BTU valve. As the BTU demand increases, the HI BTU valve opens to quickly meet the change in demand. If the demand remains high then the I-only controller gradually adds low BTU fuel and the high BTU fuel flow returns to the value specified by the I-only controller setpoint. Over time the use of high BTU fuel to satisfy steam demand is minimized. Long term, unless changes in steam demand are frequent, most of the BTU demand is met using the low BTU fuel.

Example – Ammonia H/N Control

As was described in Chapter 10, air and natural gas are the primary feed-stock inputs to an ammonia plant. The air supply to the process is provided by a large, steam turbine driven compressor. Gas is supplied to the plant through a large pipeline. The gas flow rate is normally set to achieve a target production rate and the air flow to the secondary reformer is adjusted to achieve a target H/N (hydrogen to nitrogen) ratio to the synthesis loop where the hydrogen and nitrogen are converted to ammonia. Because of the size and design of the air compressor turbine drive, small changes in the air flow rate are difficult to make through adjustment of the compressor speed. A vent valve is often installed after the compressor to allow fine adjustments in the air flow from the compressor.

The maximum flow capacity of the vent valve is small compared to the air flow through the air compressor. Valve position control may be used to precisely adjust air flow to the ammonia plant as is illustrated in Figure 13-39.

The H/N ratio controller provides a total air demand output to the air flow controller. The output of the air flow controller goes directly to the vent valve. Also, the implied valve position of the vent valve is the input to an I-only controller that adjusts the speed of the air compressor. The setpoint of the I-only controller is set to 50% to allow the compressor speed to be slowly adjusted to long term to maintain the small vent valve at 50%. Thus, the air flow can be rapidly increased or decreased using the vent valve. As the process conditions change, the vent valve may need to open or close to respond to a change in air demand. If the vent valve position deviates from 50% for a sustained period of time, the I-only controller gradually adjusts the compressor speed to allow the vent valve to return to 50%. Through the use of valve position control, the control system is able to precisely respond to the total air demand over a wide range.

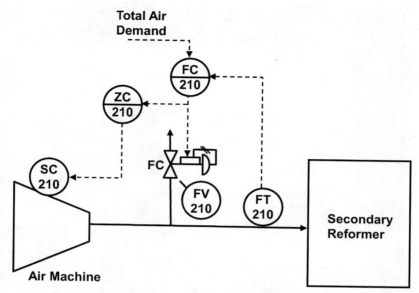

Figure 13-39. Example – Secondary Reformer Air Demand

Implementation

When implementing valve position control, both the feedback controller and the I-only controller may be implemented using a PID block. As indicated above, the output of the controller used for feedback control would be connected to the vent valve and would also be the input of the I-only controller. The blocks and connections that would be required for valve position control are illustrated in Figure 13-40.

In some control systems, the structure parameter of the PID block may be configured to select I-only control as illustrated in Figure 13-41.

When you are commissioning a valve position control strategy, the I-only controller should initially be placed in Manual. This allows the feedback controller to be tuned as a single-loop controller. Once the feedback controller is tuned, then the I-only reset should be set to be much slower. For example, if the reset of the feedback controller is 5 seconds per repeat, then the reset of the I-only controller might be set to a value that is 5–10 times larger, that is, 25–50 seconds per repeat. Also, the I-only controller setpoint should be set to the Normal position of the feedback controller output. At this point, the I-only controller may be placed in Automatic mode. Step changes in the feedback controller setpoint should be made to observe the response of each valve. If necessary, the I-only reset may be adjusted to be slower to reduce any interaction between the feedback controller and the I-only controller. When the mode of the feedback control is switched to manual, then the I-only controller will automatically suspend reset action since the feedback controller OUT limit status is constant.

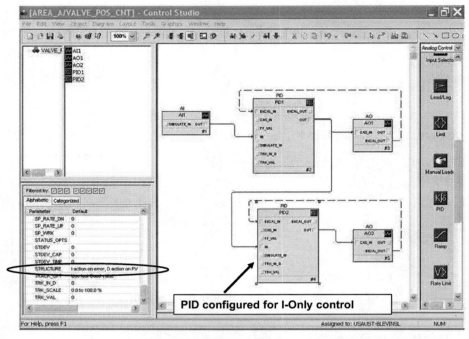

Figure 13-40. Valve Position Control

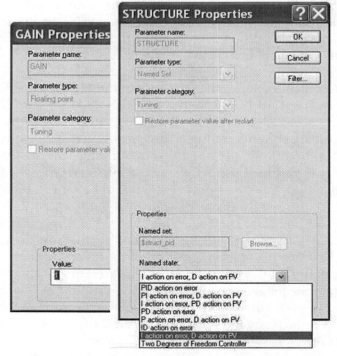

Figure 13-41. Configuring PID for I-only Control

Workshop – Valve Position Control

The exercise for this workshop is based on a simple process example where a small valve and a large valve may be used to adjust the total flow to the process as shown in figure 13-42. Valve position control is implemented as illustrated below. See the Appendix for directions on using the web-based interface.

<u>Step 1</u>: Using the web-based interface, open the Valve Position Control workshop. In the workspace shown in Figure 13-43, change the mode of the flow controller to Auto.

<u>Step 2</u>: Change the flow controller SP (Setpoint) over the following range: 40, 50, and 60. Observe the change in the two outputs. Why is the small valve maintained at 50%?

<u>Step 3</u>: Change the SP of the valve position controller and observe the response.

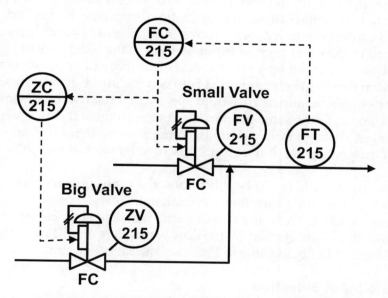

Figure 13-42. Process for Valve Position Control Workshop

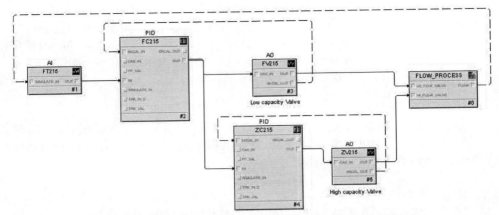

Figure 13-43. Valve Position Control Workshop Module

13.4.3   Ratio Control

For many processes, such as blending and boiler combustion, a key objective is to maintain the flow rates of two process streams in some proportion to one another. In such cases, ratio control may be applied. When ratio control is applied, one process input, the dependent input, is proportioned to the other process input, known as the independent input. The independent input may be a process measurement or its setpoint. The proportion that is to be maintained between the inputs is known as the ratio. For example, a ratio of 1:1 would specify that the two inputs are to be maintained in the same proportion. As the value of the independent input changes, through ratio control the other process input is changed to maintain the proportion of the inputs specified by the ratio setpoint.

In nearly all ratio control applications, the ratio controller sets the setpoints of the flow controllers rather than the valve positions, as illustrated in Figure 13-44. Thus, any non-linearity installed characteristics associated with valves is addressed by the flow controllers and has no impact on the ratio controller being able to maintain the ratio setpoint.

Ratio Input Selection

When ratio control is implemented, the control designer has the choice of which input to select as the independent input. In many cases, the flow that is normally the largest is selected as the independent input since this input may be used to set process throughput. If a flow controller is maintaining close control and its setpoint is representative of the process input, the designer may base the ratio on the flow controller setpoint rather than the flow measurement, which may be noisy. In this case, the setpoint value of the independent loop would be the input to the RATIO block. The output of the RATIO block would set the dependent loop setpoint equal to the ratio setpoint multiplied by the independent loop setpoint.

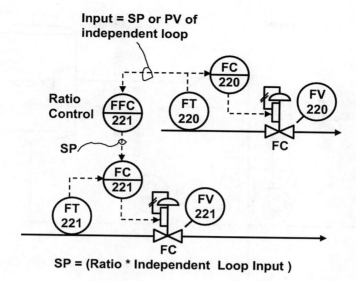

Figure 13-44. Basis – Ratio Control

As an alternative, the designer could base the ratio control on the flow measurement associated with the independent input. In this case, the dependent loop setpoint would be calculated by the RATIO block as the ratio setpoint multiplied by the independent loop measurement. Any noise in the independent flow measurement would cause changes to be made in the dependent loop setpoint.

In most cases, the independent input measurement, not the setpoint, is used as the input to the RATIO block. The reason is that for some operating conditions, the independent loop output may become saturated. If this happens, the setpoint cannot be maintained. If the ratio control is based on the independent input setpoint then the specified ratio would not be maintained. In a blending operation, such a mistake could cause the wrong quantity of a material to be added to the process.

Example – Reactor

Within a reactor, two or more feed streams are mixed and allowed to chemically react. To achieve the desired composition in the reactor outlet stream, the flow rates of the feed streams must be maintained in a fixed proportion. The feed stream flow rates are set using ratio control. This ensures that a fixed proportion is maintained when changes are made in the reactor throughput.

For example, the catalyst feed stream may be automatically adjusted to maintain a fixed ratio to the main feed stream as illustrated in Figure 13-45.

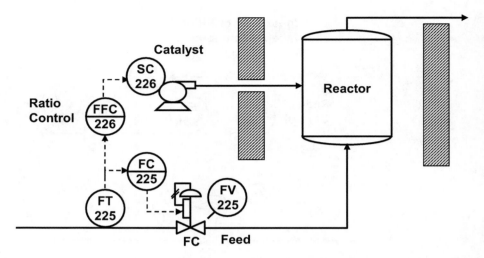

Figure 13-45. Example – Reactor Feed Ratio Control

The catalyst flow rate to the reactor is the dependent process input and is set by adjusting the speed of a positive displacement pump. The ratio controller output is used to set the catalyst pump speed. The independent process input to the reactor is the main feedstock flow. A measurement of the feedstock flow rate, rather than the flow controller setpoint, is used as the input to the ratio controller. In this application, the selection of the flow rate measurement rather than setpoint is important for correct operation since the flow controller may not be able to maintain setpoint when the output becomes limited because of a drop in supply pressure to the valve.

Example – Blending

Multiple feed streams may be mixed in a blender to achieve a target property such as ingredient concentration or composition in the blender outlet stream. When different products are made in the blender, the feed stream ratios must be changed to achieve a different target property. Also, if the composition of one or more of the feed streams changes, then the ratio must be changed to continue to achieve the target property in the outlet stream.

In some cases, the target property may be measured using an on-line analyzer. If such an on-line measurement is available, then the adjustment of the ratio setpoint may be automated using a feedback controller to manipulate the ratio setpoint as illustrated in Figure 13-46.

From the perspective of the feedback controller, the ratio controller and the flow loops controlling the feed stream flow rates are part of the equipment that make up the blender process. Once the flow controllers and

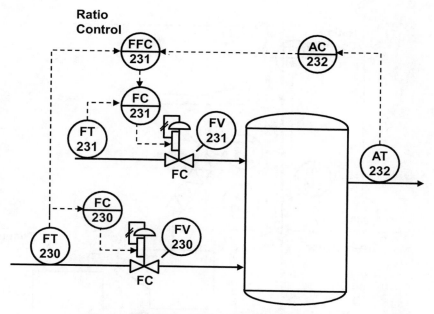

Figure 13-46. Automatic Ratio Adjustment

ratio controller are commissioned, the feedback controller operates like a single-loop controller. Since the feedback controller output sets the ratio controller setpoint, the ratio controller mode must be Cascade to allow the feedback controller to be commissioned.

Example – Slaker Lime Feed

The slaker process is used in the pulp and paper industry to convert the sodium carbonate in a green liquor stream to white liquor, sodium hydroxide, through the addition of powdered lime, calcium oxide. This reaction is optimized when the ratio of calcium oxide to sodium carbonate is maintained at a specific value. The reaction that begins in a vessel for mixing lime and green liquor, known as a slaker, is driven to completion in holding vessels, known as causticizers, which are located downstream of the slaker. Ratio control is typically used to automate the adjustment of the lime flow based on the green liquor flow input. However, the density and chemical activity of the lime and the concentration of the sodium carbonate in the green liquor may vary considerably over time. To automatically account for such variations in the feed streams, a conductivity measurement located after the causticizers is used to determine the white liquor concentration. A feedback controller is used to maintain the conductivity measurement at setpoint by regulating the ratio controller setpoint as illustrated in Figure 13-47.

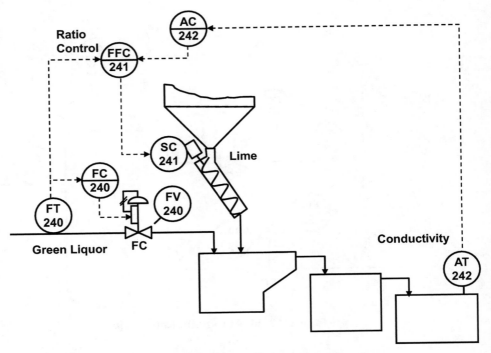

Figure 13-47. Example – Slaker Control

The lime feed to the slaker is regulated by adjusting the speed of a screw feeder. So, the output of the ratio controller is used to adjust the screw feeder speed. As the density of the lime changes, the weight of lime feed to the slaker for a given feeder speed will vary. Also, the lime reactivity and purity will impact the chemical reaction with green liquor. Such changes in the lime feed and lime composition will be reflected in the conductivity measurement and the conductivity controller will change the ratio set-point to compensate for these changes.

The green liquor flow rate is the independent input to the slaker. The flow rate is normally set by the plant operator based on the production rate and green liquor inventory.

Implementation
A typical implementation of ratio control is illustrated in Figure 13-48. The ratio block is used to implement ratio control. The input to the ratio block is the measurement or setpoint of the independent flow input to the process. This independent measurement is also known as the process "wild flow" input. In the RATIO block, this input is multiplied by the ratio to determine the setpoint of the dependent flow input to the process.

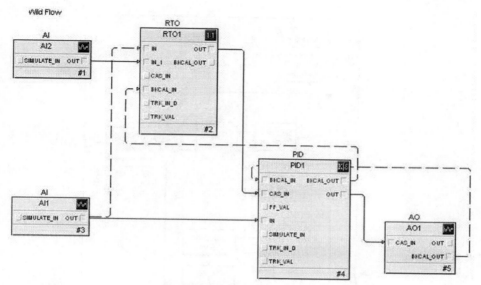

Figure 13-48. Ratio Control

The independent flow measurement is connected to IN_1 of the RATIO block. Also, the dependent flow measurement is connected to IN to allow the actual ratio (defined below) to be shown as the process variable, PV, of the RATIO block. The back calculation connection between the RATIO block and the downstream block is used to provide bumpless transfer when the downstream block is changed to Cascade mode. If the RATIO block has been configured for the setpoint to track the actual ratio, then the calculated output should match the setpoint of the downstream block. Otherwise, to provide bumpless transfer the difference between the calculated output and the setpoint of the downstream block is set as a bias that is added to the calculated output. The bias is then driven to zero at a constant rate in the time defined by the BAL_TIME parameter.

Under normal operating conditions, the RATIO block multiplies the independent input flow provided on IN_1 by the RATIO block setpoint to provide an output to the downstream block as illustrated in Figure 13-49. The actual ratio is calculated by the block using the IN and IN_1 inputs. Block alarm limits based on the actual ratio and the setpoint may be configured.

The deviation alarm could be used to alert the operator when the ratio setpoint cannot be maintained. For example, if the downstream block output is limited, then the setpoint provided by the RATIO block may not be maintained. Since the flow measurement is used to calculate the actual ratio, this limited condition would be indicated by the ratio PV not matching the setpoint. If the deviation between the setpoint and PV exceeds the deviation alarm limit, an alarm can be generated by the RATIO block.

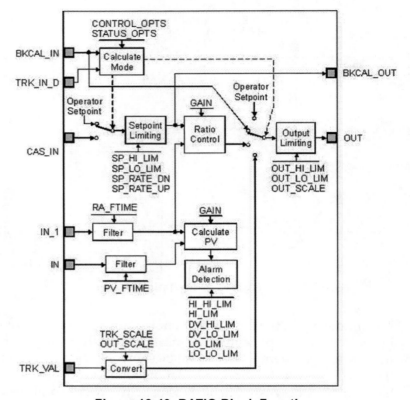

Figure 13-49. RATIO Block Function

Workshop – Ratio Control

In this workshop, a blender is used to blend two streams of different concentrations. The percent solids concentration exiting the blender is measured on-line and is the input to a concentration controller. The concentration controller output adjusts the setpoint of the ratio controller associated with the main flow and blend as shown in Figure 13-50. Changes in the concentration controller setpoint cause the inlet flow ratio to change to the value needed to achieve the target outlet concentration. See the Appendix for directions on using the web-based interface.

Step 1: Using the web-based interface, open the Ratio Control workshop. In the workspace shown in Figure 13-51, change the mode of the RATIO block to Auto.

Step 2: Change the Ratio SP (setpoint) over the following range: 0.3, 0.5, and 0.8. Observe the change in the blend flow and the process outlet concentration. Then set the Ratio SP to 0.5 and wait for the concentration to settle to a steady value.

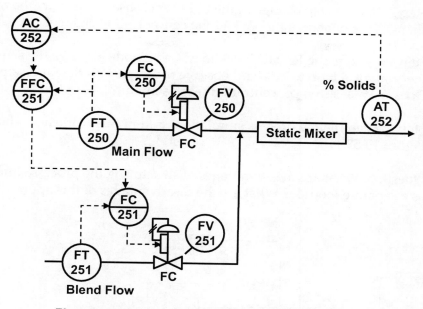

Figure 13-50. Process for Ratio Control Workshop

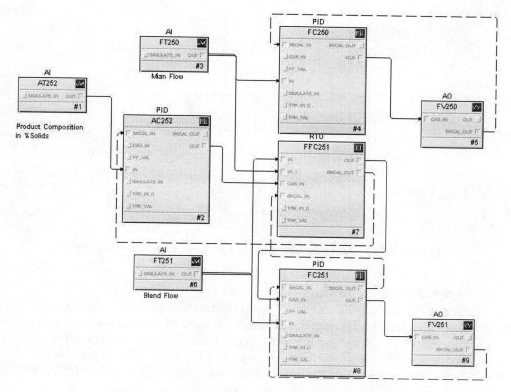

Figure 13-51. Ratio Control Workshop Module

<u>Step 3</u>: Make a step change in the FEED and observe the way the ratio changes the dependent loop. Did the concentration change?

<u>Step 4</u>: Change the RATIO block to cascade mode. Change the setpoint of the analytical loop to 40% and observe the impact on the ratio setpoint. Does the measured concentration reach setpoint?

<u>Step 5</u>: Examine the process simulation used in this workshop, shown in Figure 13-52.

Question: What are the advantages of structuring the process simulation as a separate module? What are the disadvantages of this approach?

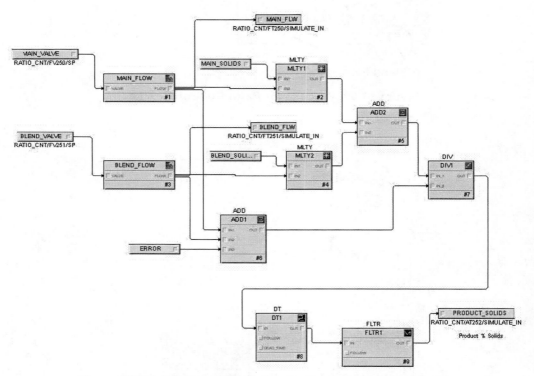

Figure 13-52. Process Simulation for Ratio Control Workshop

Workshop Discussion

In Automatic mode, as the setpoint of the inlet flow stream was changed, the PV of the RATIO block reflected the ratio of the inlet stream flow rates. Since the option for the setpoint to track the PV was enabled in the RATIO block, the SP automatically followed the PV when the dependent flow loop was not in Cascade mode. This bumpless transfer should have been observed when the mode of the dependent flow loop was changed to Cascade.

Since an on-line measurement of concentration was available, it was possible to automatically adjust the ratio setpoint to achieve the desired outlet concentration. If the concentration could only be measured using an off-line lab test, then the operator would have to adjust the ratio setpoint manually based on the lab test.

The workshop exercise was designed to show the effectiveness of ratio control. Little or no change in the outlet concentration should have been observed as the flow rate of the independent flow stream was changed.

In production, the flow rate of the independent flow stream in a ratio control strategy is normally adjusted to meet the plant production schedule.

The advantages of structuring the process simulation as a separate module are addressed in Chapter 15.

14

Model Predictive Control

The techniques that are most often used to address process control requirements have a long history of use in the process industry. Some of the first control systems used in centralized control rooms in the mid-1900s were based on pneumatic field transmitters and pneumatic controllers. Even in these early control systems, there are examples in which cascade control was implemented. Using pneumatic computing elements, it was also possible to achieve feedforward control. The introduction of panel-based electronic controllers and field devices addressed many of the distance limitations associated with pneumatic transmission lines. Also, the electronic computing elements made it easier to implement feedforward and override control strategies.

When digital distributed control systems (DCS) were introduced in the late 1970s, control functionality was just a digital implementation of techniques that were common in an electronic, panel-based control system. So, it caused quite a stir when Cutler and Ramaker presented a paper in 1980 [1] providing information on a new technique known as Dynamic Matrix Control. Shell Oil had developed and deployed this technique for the control of large interactive multiple input-multiple output (MIMO) processes such as refinery distillation columns. This work by Shell was the first version of what is commonly referred to today as Model Predictive Control (MPC).

Model Predictive Control has been installed in thousands of plants since the initial installations at Shell. The multi-variable constraint handling capability of model predictive control may often be used to increase a plant's production rate. However, the computation power and memory needed to implement model predictive control may require a computer that is layered on top of an older distributed control system, so the application of MPC has traditionally been limited to large, high throughput processes where an increase in throughput can justify the high cost of installing, commissioning, and maintaining an MPC strategy.

In a modern DCS, advances in the processors and memory of the controller make it possible to embed MPC in the controller. [2] If the DCS controller supports MPC, then it is possible to apply MPC to small processes that have historically been controlled using multi-loop techniques. Also, MPC may be used to more effectively control processes that are dominated by deadtime and difficult dynamics such as inverse response than is possible with PID. This chapter addresses the application of MPC to small processes and to processes that are dominated by deadtime and difficult dynamics, and the benefits of this approach versus traditional control techniques based on PID.

14.1 MPC Replacement of PID

Model Predictive Control may be used for the control of single input-single output (SISO), as well as multiple input-multiple output (MIMO) processes. In its simplest form, MPC may be used to address a SISO process. For example, MPC can be used to provide an effective means of controlling a SISO process that may be described as "deadtime dominant" when the process deadtime is equal to or greater than the process time constant. The control performance achieved may not be satisfactory when PID feedback control is applied to a deadtime-dominant process. As the ratio of deadtime to time constant increases, the proportional and reset gains must be decreased for stability. As a result, the control response to changes in setpoint and load disturbances is often slower than what is required for best process operation. In such cases, control performance may be improved by replacing PID feedback control with Model Predictive Control as illustrated in Figure 14-1.

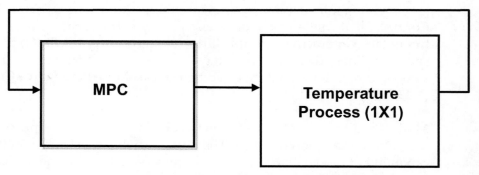

Figure 14-1. MPC – Addressing Difficult Dynamics

This improvement over PID feedback control performance is possible because the MPC algorithm is generated based on process response to a step change in process inputs, known as the step response model, rather than the algorithm being pre-defined as is the case with PID algorithm. When a change is made in an input of a process characterized by dead-time, unlike the PID algorithm the MPC algorithm is aware that this change will not be immediately reflected in the process. This awareness of the process response allows the MPC algorithm to do a better job of controlling deadtime-dominant processes.

Implementing model predictive control of a SISO process is as simple as using PID control. The MPC block simply replaces the PID block in the control module as illustrated in Figure 14-2. The connection of the analog input and output blocks to the MPC block is done in exactly the same way. However, when using MPC, the controller tuning is addressed in a different manner. For example, the MPC block does not contain parameters for proportional gain, integral gain, and derivative gain. Instead, the identified step response, known as the step response model, is used to generate the MPC control algorithm.

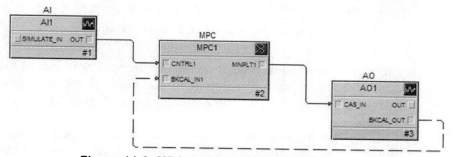

Figure 14-2. MPC Configuration for Single Loop

14.2 Commissioning MPC

When MPC is supported in a control system, a software application is provided to automatically identify the process step response model. Similar to the tuning applications provided to tune the PID, the MPC application is designed to allow the process to be automatically tested. With the push of a Test button, the step response model is automatically identified and the MPC algorithm is generated and automatically transferred to the MPC block used for control. The identified step response may be viewed as illustrated in Figure 14-3.

Before initiating testing, the user may specify the amount the manipulated parameter is to be changed during testing. Also, it is necessary for the user

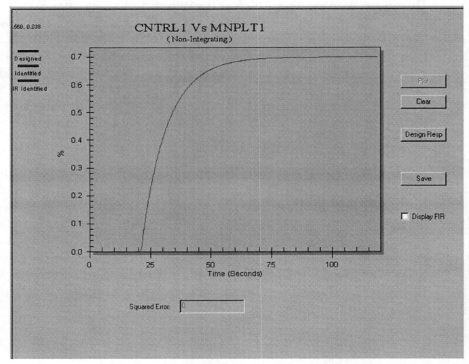

Figure 14-3. Identified Process Step Response

to input their best guess, based on observation, of how long it takes the process to fully respond to a change in the manipulated parameter, that is, the process time to steady state as illustrated in Figure 14-4.

For example, if the user should say, "It is pretty well settled out after two minutes," then 120 seconds would be entered as the time to steady state. The duration of the automated test is based on the estimated time to steady state and also determines the maximum time duration of pulses generated during testing. When testing is initiated, the controller output will be automatically changed to generate a series of pulses the duration of which changes in a pseudo random fashion to allow the process response to be observed and collected for model identification.

> **Pseudo Random** - A series of changes that appears to be random but are in fact generated according to some prearranged sequence.

After testing is complete and Autogenerate is selected by the user, the manner in which the process input and output varied during the test are automatically analyzed and the step response model and MPC controller are generated.

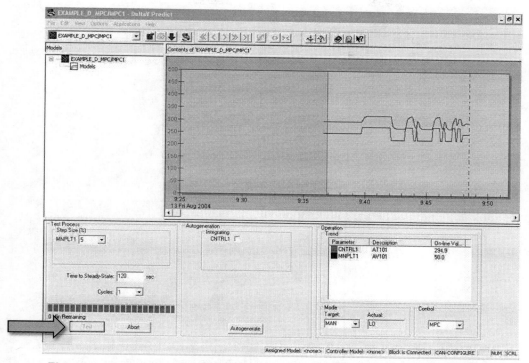

Figure 14-4. Automated Process Testing to Identify the Process Model

In general, MPC may also be used to control processes having multiple inputs and multiple outputs. When MPC is applied to a process with multiple inputs and outputs, the automated testing would introduce changes in all manipulated process inputs and data would be collection of all process inputs and outputs to support model identification. The display of the step response model is designed to show the step response of each output for a step change in each input where all other inputs are maintained constant. For example, if the MPC block supports a process with as many as four manipulated and four disturbance inputs, and four controlled parameters and four constraint outputs, then the identified step response display for a single input and single output process might appear as illustrated in Figure 14-5.

For a process with only one input and one output, one step response will be shown in the display as illustrated in Figure 14-5. A larger view of the step response that is displayed may be obtained by clicking on that step response.

When the MPC block is configured for multiple process inputs and/or outputs, then for each manipulated or disturbance input, the impact of a 1% change in this input (with all other inputs constant) on each process

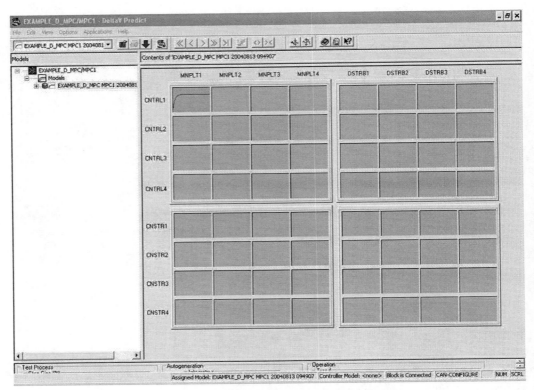

Figure 14-5. Step Response Model

output is shown. When an MPC block is used to control a process that has eight process inputs and eight process outputs, the step response model would contain 64 step responses. In such a case, the process might be referred to as an 8x8 process to indicate the number of process inputs X number of process outputs.

When a step response model has been identified by testing the process, a natural question to ask is, "How accurate is the identified step response model prediction?" To validate the identified step response model, the changes in the manipulated and disturbance input that occurred during testing can be passed through the model. The response calculated using the model can then be plotted against the actual process output response as illustrated in Figure 14-6.

In this example, for the process output selected in the right-hand pane of the display, there is little noticeable difference in the calculated and the actual value of the process output. This indicates that the model identified by the MPC application accurately predicts the process output response to changes in the process input.

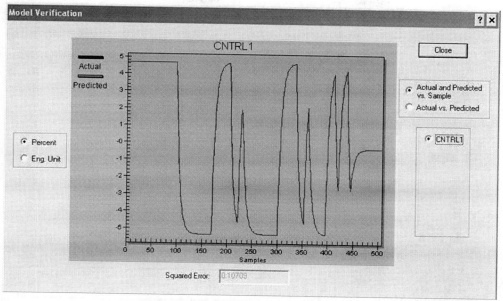

Figure 14-6. Verification of Identified Model

Similar to the PID tuning application, the MPC application may be designed to allow the control response to be observed before downloading the module that contains the MPC block based on the identified step response model. By selecting Simulate, it is possible to view how the process will respond to a setpoint change. One of the unique features of model predictive control is that since the controller is generated from the process model, the MPC block internally uses the model to predict process outputs based on changes in the process inputs. This predicted response may be shown in the operator interface to an MPC block and is also shown when the MPC block is viewed in the simulation environment. Such a prediction is especially valuable for operator guidance when one or more manipulated variables are in MAN (Manual; i.e., not being manipulated by MPC). In the example shown in Figure 14-7, the predicted response of the process is shown in the lighter area on the right side of the trend plot.

If the process time constant and deadtime are very large, then a change in a process input may not be reflected in the process response for many minutes or even, in some cases, many hours. Thus, the MPC simulation environment may be designed to allow the process response to be displayed much faster than real time. In the example application shown in Figure 14-7, the Slew button in the upper right corner of the screen may be used to view the process response as much as 180 times faster than real time.

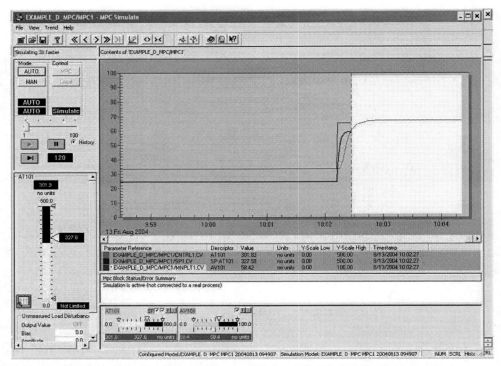

Figure 14-7. Testing of Control Using Simulated Environment

To allow MPC control to be viewed and worked with in exactly the same manner as with PID control, some manufacturers provide pre-built dynamos (dynamic elements) that may be included in the operator display. From the operator's perspective, single-loop MPC control may then be used in exactly the same manner as single-loop PID control. Also, to support an engineer who is working with an MPC application, pre-built dynamos may be supplied by the manufacturer that are very similar to the interface provided for simulation, as illustrated in Figure 14-8, but with the speedup capability removed.

An engineer who is used to working in the MPC simulation environment may use this on-line interface to view and work with an MPC block that is controlling the process.

14.3 MPC Replacement for PID with Feedforward

If a disturbance to a process can be measured, then this input may be incorporated into the MPC as illustrated in Figure 14-9. In this way, MPC control may be used to address applications that are traditionally addressed by adding a feedforward input to the PID. When a measured disturbance is connected as an input to the MCP block, the MPC applica-

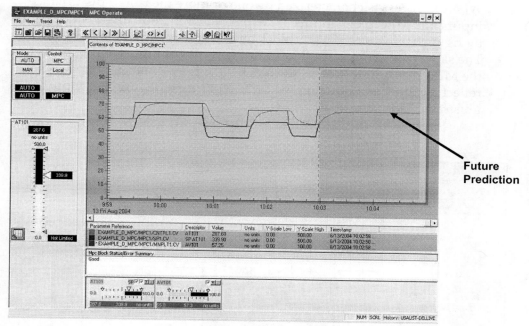

Figure 14-8. Operator Interface to MPC

tion may be used to automatically identify the impact of the disturbance on the process output. By including the measured disturbance as an input, the MPC block can automatically compensate for the disturbance and thus do a better job of maintaining the controlled parameter at setpoint.

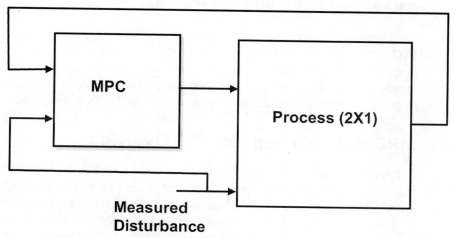

Figure 14-9. MPC Replacement for PID with Feedforward

When a process is characterized by one or more measured disturbance inputs, the MPC block inputs are expanded to include disturbance inputs by right clicking on the MPC block and selecting Extensible Parameters. The same procedure is used to add other types of inputs and outputs to the MPC block. In response, the block view is automatically updated to reflect a connection point on the MPC block for a disturbance input, as illustrated in Figure 14-10.

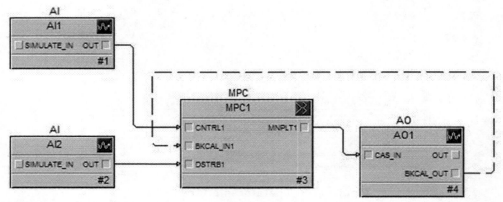

Figure 14-10. Example MPC Implementation for One Measured Disturbance Input

The default configuration information associated with MPC block inputs or outputs may be viewed or changed by right clicking on the block and choosing Properties. In response, a dialog box will be presented for viewing and changing the associated configuration; for example, setpoint high and low limit or output low and high limit.

MPC configuration typically involves significantly fewer parameters than PID configuration of an equivalent control strategy. Since the MPC block is generated from the step response model, corrections are automatically made for any differences in the controlled parameter response to changes in the disturbance inputs and the manipulated process inputs.

14.4 MPC Replacement for PID Override

When a process output is a constraint parameter in PID feedback control, this constraint parameter measurement is simply added as an input to the MPC block as illustrated in Figure 14-11. The design and implementation of an override control strategy using an MPC block is thus much simpler than implementing override control using two PID blocks and a control selector block.

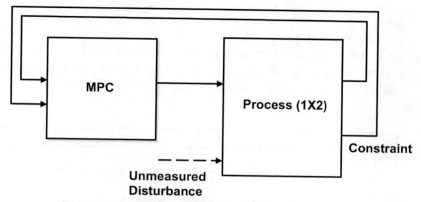

Figure 14-11. MPC Replacement for PID Override

When a measured constraint is included as an input to the MPC block, the step response of the constraint parameter to changes in the process input(s) is automatically identified when using the MPC application to test the process and to generate the step response model. A constraint parameter is added to the MPC block by right clicking on the block and selecting Extensible Parameters. The view of the block is updated to include the constraint input as illustrated in Figure 14-12.

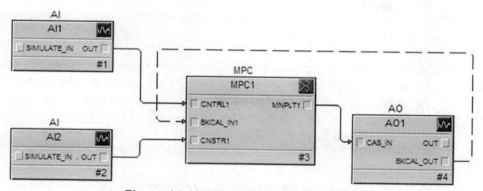

Figure 14-12. MPC Constraint Control

Since the process step response is used to generate the MPC controller, the response of constraint parameters and controlled parameters to changes in the process inputs are automatically taken into account to minimize varia-tion from setpoint while observing the constraint limits.

14.5 Using MPC to Address Process Interactions

As we saw in Chapter 12, when a process is characterized by multiple manipulated process inputs and multiple controlled process outputs, there is a potential for process interaction, that is, a change in one manipulated input may impact multiple controlled outputs. When the resulting interaction between control loops is significant, detuning the PID controller is the most common way to deal with it. When MPC is used, the interaction of the manipulated inputs and controlled outputs may be addressed by one MPC block as illustrated in Figure 14-13.

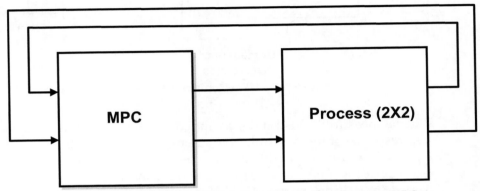

Figure 14-13. Addressing Process Interaction Using MPC

Since the impact of each manipulated input parameter on each controlled output parameter is identified by the step response model used in MPC block generation, any interactions are automatically compensated for by the MPC block.

Manipulated process inputs and controlled process outputs are added to an MPC block by right clicking on the block and selecting Extensible Parameters. In response, the view of the block will be updated to show the added inputs and outputs. When two manipulated inputs and two controlled outputs have been configured, the MPC block and its connections to associated input and output blocks are as illustrated in Figure 14-14.

A typical step response model for a process with two manipulated inputs and two controlled outputs is illustrated in Figure 14-15. In this example, manipulated input 1 has a much larger impact on controlled output 1 than on controlled output 2 while manipulated input 2 has a larger impact on controlled output 2. Significant process interaction exists since the step response model shows that a change in either input has an impact on both controlled outputs.

Since the step response model is used to generate the MPC block, these interactions are automatically taken into account. Thus, each controlled parameter may be independently maintained at setpoint without interaction.

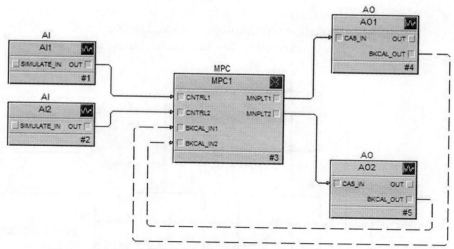

Figure 14-14. MPC Implementation for Interactive Process

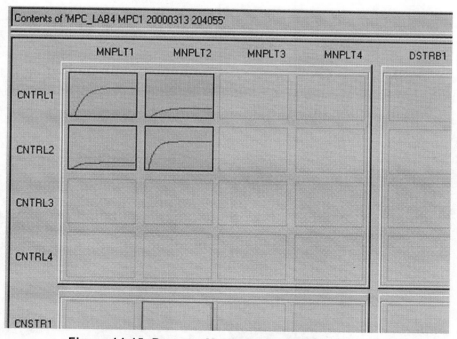

Figure 14-15. Process Model for Interactive Process

14.6 Layering MPC onto an Existing Strategy

In many cases, the use of MPC in place of PID control is a new concept to a control engineer who has designed and commissioned control systems and used traditional control strategies based on PID. An easy way to initially learn about MPC and to gain experience commissioning MPC blocks is to layer MPC blocks on top of PID-based traditional control strategies, as illustrated in Figure 14-16. The output of the MPC block is connected to the RCAS_IN of the analog output block. The RCAS_OUT of the analog output block is connected to the back calculation input of the MPC block.

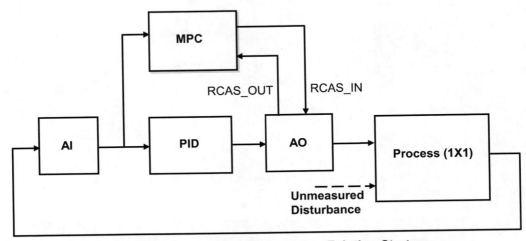

Figure 14-16. Layering MPC onto an Existing Strategy

In this example, the controlled parameter is wired as an input to both the PID block and the MPC block by exposing the RCAS_IN and RCAS_OUT parameters of the AO block, as illustrated in Figure 14-17.

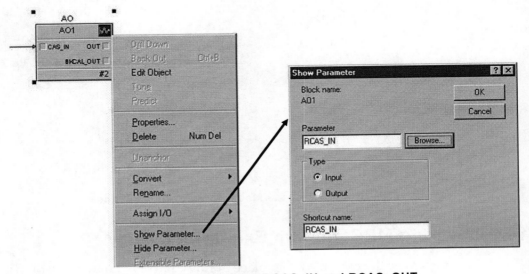

Figure 14-17. Exposing RCAS_IN and RCAS_OUT

Once the RCAS_IN and RCAS_OUT parameters of the AO block are exposed, they may be used for connections to the MPC block output and back calculation input as illustrated in Figure 14-18.

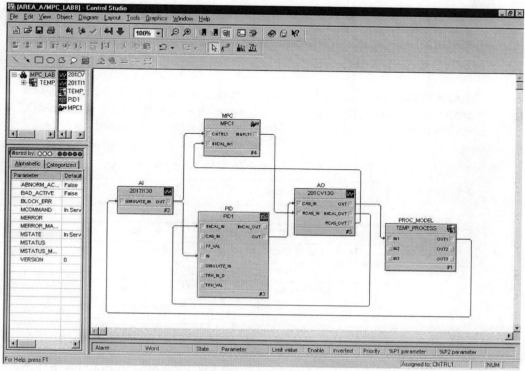

Figure 14-18. Implementing MPC onto an Existing Strategy

By changing the AO block mode to RCas, the engineer can allow the MPC block to control the process. When the AO mode is switched back to Cascade, the PID block may be used to control the process. Switching between Cascade and RCas mode is bumpless since the PID and MPC blocks both have back calculation connections.

The control system may initially be commissioned using PID control since this is what the engineer and operator typically have the most experience using to operate the plant. However, as time permits, the MPC block may be commissioned and gradually used more and more in the control of the process. When the distributed control system supports the layering of MPC onto an existing strategy, then either option, PID or MPC, is always available to operate the plant.

As mentioned earlier, MPC may be used to address situations where control performance using PID is limited. For example, if the process deadtime is significant compared to the process time constant, then better control performance may be achieved using MPC. If there are strong process interactions, then this is another reason to use MPC rather than PID. Also, if a process is characterized by many constraint parameters, then the implementation and commissioning of the control system will be much faster and easier using MPC rather than multiple PID blocks and one or more control selectors.

14.7 MPC Applications

Model Predictive Control has been applied to many different process applications. Table 14-1 lists a small sample of applications where MPC has been used to replace traditional control strategies based on PID.

Table 14-1. Example MPC Applications

Control Application	Industry
Evaporator	Chemical
Lime Kiln	Pulp & Paper
Pipeline Gas Blending	Power
Bleach Plant	Pulp & Paper
pH Control	Pulp & Paper
Distillation Column	Pharmaceutical

As has been noted, the process industry has been slow to use MPC to replace PID in smaller process applications. One reason may be that there is often not enough time in a project to try something new. Also, only the latest distributed control systems support MPC at the controller level. However, little by little, plant by plant, the use of MPC for small applications is gaining momentum. When MPC is available as a function block in the control system, implementation is much easier and less costly than by using a computer layered on top of the control system.

14.8 Workshop – Model Predictive Control

The control performance of MPC versus PID for override control is demonstrated in this workshop. To allow this comparison of PID and MPC, in this workship an MPC block is layered on top of a PID override control. By changing the mode of the AO block from Cas to RCas it is possible to switch between PID and MPC control of the process. See the Appendix for directions on using the web-based interface.

<u>Step 1</u>: In the Model Predictive Control workspace, observe the way MPC has been layered on top of PID override control. With the AO block in Auto, change the AO block setpoint by a step and observe the response.

<u>Step 2</u>: With the AO block in Cascade mode, change the disturbance input and observe the response of the PID override control.

<u>Step 3</u>: Switch the AO block to RCas mode, change the disturbance input, and observe the control response of the MPC.

Question: Was the observed response equivalent (PID versus MPC)?

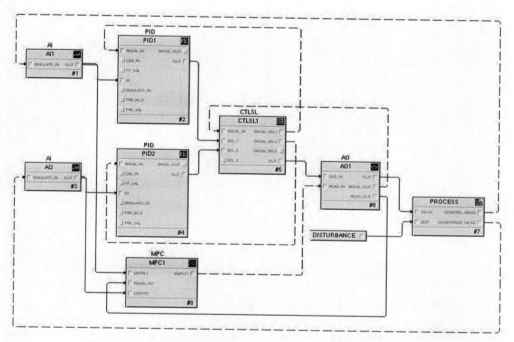

Figure 14-19. Module MPC Workshop – PID versus MPC

References

1.   Cutler, C. and Ramaker, B. "Dynamic Matrix Control—A Computer Control Algorithm," Proc. Joint Automatic Control Conference, San Francisco, CA, 1980.

2.   Blevins, T., McMillan, G., Wojsznis, W., and Brown, M. *Advanced Control Unleashed: Plant Performance Management for Optimum Benefit*. Research Triangle Park: ISA, 2003 (ISBN: 978-1-55617-815-3).

15

Process Simulation

Dynamic process simulations are used in the modules provided for the workshop exercises on process characterization and single-loop and multi-loop control techniques. The process simulation was embedded in many of the modules as a composite block. Also, as discussed in the ratio control workshop in Chapter 13, the simulation may be created in another module.

Using this capability to create the process simulation in another module, process simulations may be added to a control system to check out or demonstrate a control strategy without modifying the modules used for control. Also, comprehensive operator training systems may be created using this capability.

Since a process simulation can be such a useful tool when working with a control system, this chapter addresses how to add process simulations to a control system using tools that exist in most control systems. We will discuss:

- Techniques for adding a process simulation to a control system
- Developing a process simulation starting with the Piping and Instrumentation Diagram (P&ID)
- Simulating process non-linearity
- Other considerations

Examples of more a complex process that may be simulated using this approach are provided in Chapter 16, Applications.

15.1 Process Simulation Techniques

In a control module, the analog output block is used to manipulate final control elements, such as valves, dampers, or variable speed drives. Since the analog output block OUT parameter represents the current or digital signal to these final control elements, this parameter is connected as an input to the process simulation. Similarly, the OUT_D parameter of discrete output blocks may be added as inputs to the process simulation. Measured and unmeasured disturbances are included in the simulation as adjustable discrete or analog parameters. These parameters are also connected as inputs to the process simulation.

The process conditions that result from the manipulated and disturbance inputs are reflected in outputs of the process simulation. For example, monitored measurements of process conditions, such as flow, pressure, temperature, level, composition, or the discrete state of a limit switch, may be easily included in the simulation. Some of these process conditions may be measured in the field using transmitters or switches and thus the simulated value can be provided to the modules that access these field values using analog and discrete input blocks. The current or digital outputs of these transmitters and switches are normally accessed in a control system using the analog and discrete input blocks. However, the normal processing of analog and discrete input blocks may be altered by enabling the Simulate parameter, as illustrated in Figure 15-1.

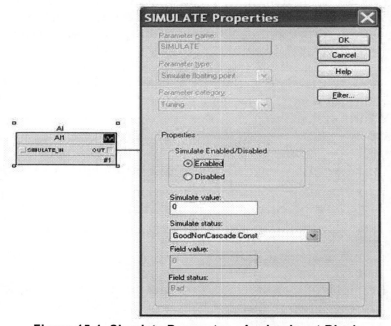

Figure 15-1. Simulate Parameter—Analog Input Block

When simulation is enabled in an input block, a Simulate parameter value is used in place of the field measurement. Thus, the simulated process output values may be used by analog and discrete input blocks in place of field measurements. Via this basic capability of the control system, process simulation may be incorporated by one of two means:

- Embedding Simulation in existing modules
- Adding Process Simulation modules

In most of the modules used in the workshop exercises, the process simulation was embedded as a composite block in the control module. It is quite fast and easy to add a composite block to an existing control system module. This approach also allows the blocks used for control and the process simulation to be observed together in the same module. In addition, the parameters of blocks in the module, or blocks that make up the composite block, may be accessed and changed through this module. For example, by double-clicking on a composite block, the blocks that make up the process simulation may be accessed and changed as illustrated in Figure 15-2.

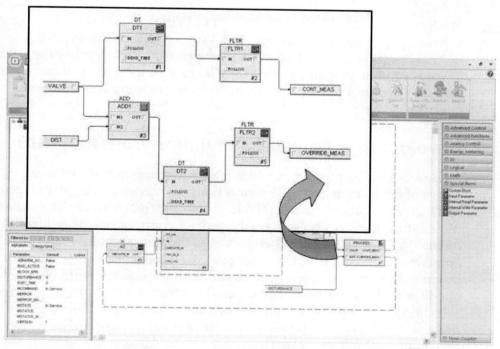

Figure 15-2. Accessing Details of Embedded Process Simulation

In some cases, from an engineering and maintenance perspective modifying existing control system modules is often not desirable. Process simulation may be useful for control system checkout and operator training but is not needed for plant operations. Duplicate modules, that is, one set of modules with simulation and one set of modules without simulation, can be created. However, updating two sets of modules to reflect changes made during checkout, commissioning and startup can complicate system maintenance. Because of this, the practice of embedding simulation in an existing module is most often used to create learning tools, such as the modules used for the workshop exercises.

The most common means of adding process simulation to a control system is to create new modules that contain only the process simulation. These simulation modules may be added without modifying existing modules. Most modern control systems include the capability for a module to read and write parameters contained in another module. Using this capability, simulation modules may read the output value of analog and discrete output blocks of an existing module. The results of the simulation modules are then written back to the simulation input parameters of the analog and discrete input blocks of the existing modules. An example of this implementation approach is illustrated in Figure 15-3.

The Simulate parameter of the I/O blocks in the existing modules may be enabled to use the process simulation. When simulation is enabled, the normal operator control displays and engineering tools to view the module operation on-line and to tune the PID may be used to view the monitoring and control functions performed by the existing control system modules.

15.2 Developing a Process Simulation from the P&ID

As detailed in Chapter 7, Control Strategy Documentation, the major pieces of equipment and their connections are shown on the plant Piping and Instrumentation Diagram (P&ID). Also, the final control elements and transmitters that are accessed by the process control system are identified on the P&ID. So, the P&ID often serves as the starting point in the development of a process simulation for control system checkout or operator training. In most cases, one process simulation module should be developed for multiple modules that support control or monitoring functions associated with a piece of equipment. Thus, a first step in developing a process simulation is to break the equipment and piping shown on the P&ID into multiple processes that may be simulated in one or more simulation modules.

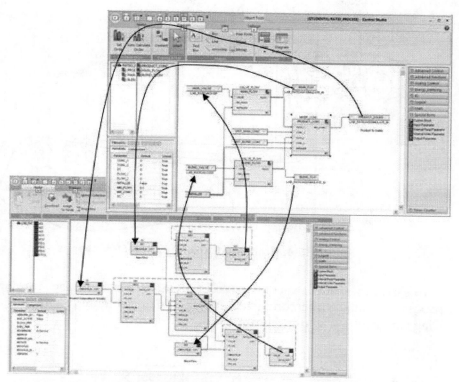

Figure 15-3. Adding Simulation Modules to a Control System

As a general rule, pieces of equipment or piping that contain transmitters and final control elements used in a single-loop or a multi-loop control strategy should be represented as one or more processes in the same simulation module. For example, the actuator position implied by an analog output block in a control module is an input to the simulated process. Any measurements in the flow path between the actuator(s) and transmitters used in control are also included as outputs of the simulated process. Unmeasured disturbances may be included as parameters of the simulation module. These disturbance parameters may be changed when you are testing a control strategy, or during operator training, to demonstrate the impact the disturbance has on the process.

Each simulated process represents a small portion of the plant equipment and interconnecting piping shown on a P&ID. The calculated output of one process simulation may be an input to another process simulation. An example of how the equipment and piping shown on the P&ID may be broken into smaller sets of equipment that represent processes is illustrated in the Recycle Tank example shown in Figure 15-4.

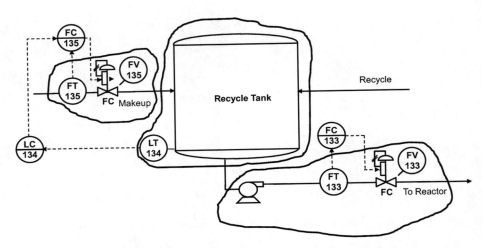

Figure 15-4. Breaking P&ID into Small Processes

In this example, the areas of the P&ID circled in black will be simulated processes that are interconnected and can be documented in a Simulation Diagram as illustrated in Figure 15-5.

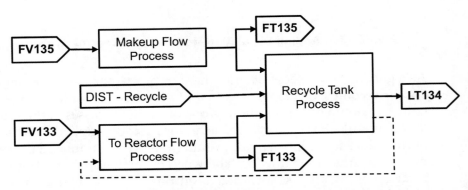

Figure 15-5. Diagram Showing Processes for Recycle Tank

Once the P&ID has been analyzed in this manner, the simulation diagram is implemented as a simulation module that references the existing control modules. The processes on the simulation diagram are implemented as composite blocks in the simulation module. Process inputs from final control elements are added as external references to the setpoints of the analog output blocks in the existing control modules. Unmeasured inputs to a process should be added as parameters. Outputs of the process simulation that represent field transmitter measurements are implemented as external references that write to the simulate parameter of the analog input blocks of the existing modules.

The simulation module for the recycle tank example is shown in Figure 15-6.

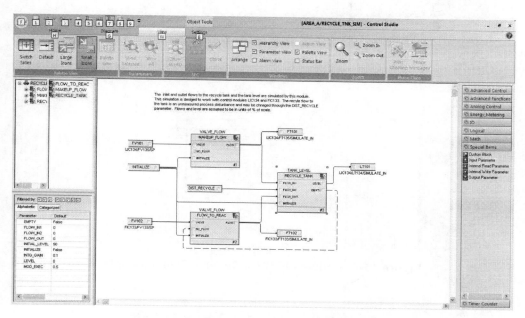

Figure 15-6. Simulation Module

For the recycle tank example, tank level control LC134 and makeup flow FC135 are implemented in one control module, LIC134, as illustrated in Figure 15-7.

The flow loop to the reactor in the recycle tank example is implemented as one module, FIC133, as illustrated in Figure 15-8.

When a control system is commissioned, the step response of a self-regulating process is often characterized as first order plus deadtime. Similarly, the deadtime and integrating gain are used to characterize the step response of an integrating process. So, it is appropriate that a process simulation be designed to duplicate the step response. For example, a filter block, a deadtime block and a multiplier block may be used to simulate the process deadtime, time constant and gain of a single input-single output (SISO) self-regulating process. Process noise may be added to the simulated process outputs using a signal generator block. For example, the FLOW_VALVE composite illustrated in Figure 15-9 uses this approach to simulate the makeup and reactor flow processes. This composite was created using these standard function blocks.

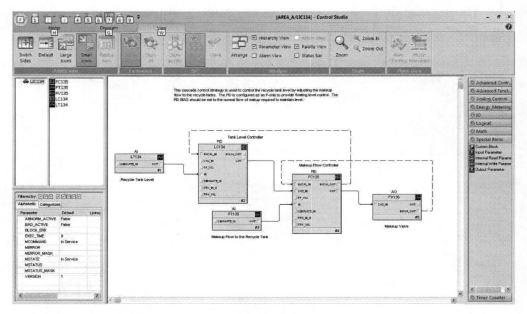

Figure 15-7. Module LIC134 – Recycle Tank Example

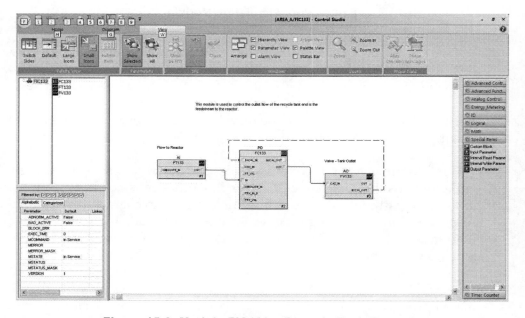

Figure 15-8. Module FIC133 – Recycle Tank Example

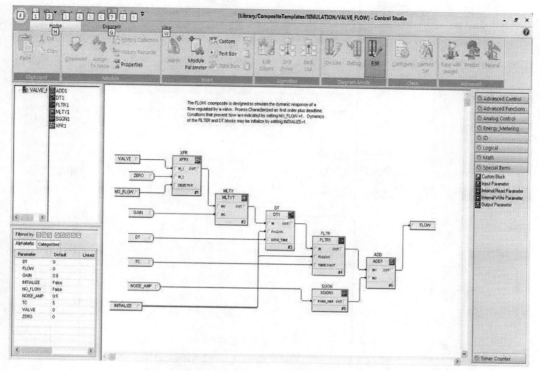

Figure 15-9. FLOW_VALVE Composite

Additional inputs may be included in the process simulation to address abnormal operating conditions. As illustrated in the FLOW_VALVE composite, a NO_FLOW input is included to allow the simulation to accurately reflect conditions that prevent flow.

Outputs may also be included in the simulation to reflect abnormal conditions. For example, the LEVEL_TANK composite used to simulate the recycle tank includes an EMPTY output. In the level simulation, this Boolean parameter transitions to a value of 1 when the level is at zero. As illustrated in the simulation module for the recycle tank shown in Figure 15-6, this EMPTY condition is used in the simulation of the flow to the reactor.

The simulation of an integrating process should take into account the module execution period and should support a varying number of inputs. Also, the calculated value for the level should be limited to the container/ vessel height. A combination of standard function blocks and a calculation block may be used to implement the tank level simulation as illustrated in Figure 15-10.

The real-time dynamic response of the simulated recycle tank and associated control modules will accurately reflect changes in the flow to the

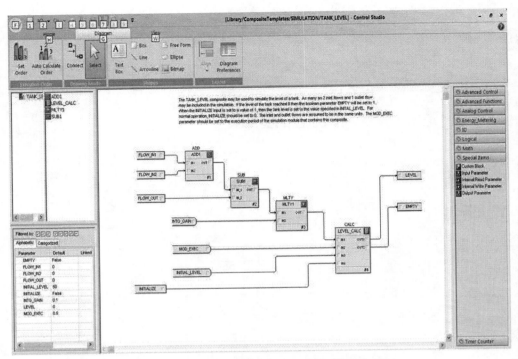

Figure 15-10. Recycle Tank Level Simulation

reactor and in the recycle tank inlet flow disturbance input that is defined in the recycle tank simulation module.

A simulated process may have multiple inputs and outputs. In the recycle tank level simulation, an identical change in inlet and outlet flow rates could be safely assumed to have the same dynamic impact on the tank level. However, in general each process input may impact each process output in a different manner. As addressed in Chapter 14, Model Predictive Control, the impact of each process input on each process output may be characterized by its step response. Process simulations can be designed to duplicate the process step. The composite blocks included in the modules used in the Multi-loop workshop exercises (Figure 15-11) are examples that show how this simulation may be implemented.

In examining the process simulation diagrams, it can be seen that process simulations used in all the multi-loop workshop exercises permit a maximum of two process inputs and two process outputs. The process simulation for these exercises could have been implemented using a process simulation that is designed to support two process inputs and two process outputs as illustrated in Figure 15-12.

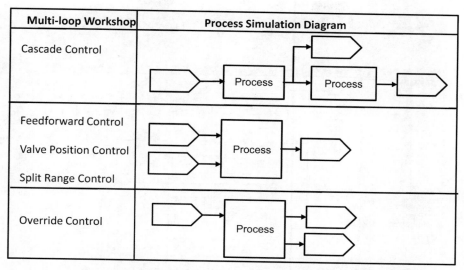

Figure 15-11. Simulation for Multi-loop Workshop

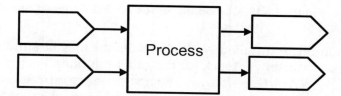

Figure 15-12. Process Simulation Diagram – 2x2 Process

The implementation that could be used to simulate processes as large as 2x2 is shown in Figure 15-13. To implement this composite block only standard function blocks are required.

The composite for simulating 2x2 processes is a generalization of the techniques used in the implementation of the recycle tank simulation composites. This approach may be easily expanded to simulate much larger processes such as distillation columns and reactors as is illustrated in Chapter 16 on Applications.

15.3 Simulating Process Non-linearity

Simulations based on duplicating the process step response may be effectively used to calculate a variety of process measurements. However, this simulation technique assumes that the process behaves in a linear fashion. When the simulation is used over a wide operating range, the step response may not accurately show the impact of process non-linearity.

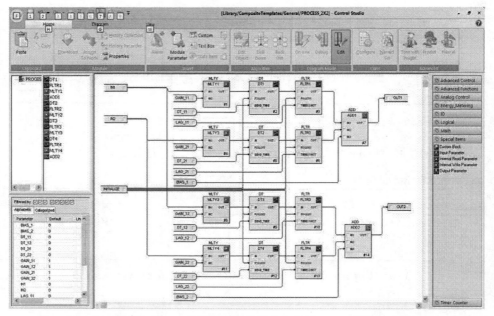

Figure 15-13. Composite Block for 2x2 Process Simulation

A non-linear installed valve characteristic may be simulated by substituting a characterizer block for the gain multiplier block in the FLOW_VALVE composite. The non-linearity introduced by an oddly-shaped tank may be simulated by including a characterizer block in the LEVEL_TNK composite to calculate the integrating gain as a function of level. In a similar fashion, the non-linear response often seen in analytic measurements, such as concentration after blending two streams or heater outlet temperature for varying inlet flow rates, may be accounted for in the process simulation.

The necessary corrections in process gain are based on input/output relationships determined by doing an energy and/or mass balance for steady-state operation. From the simulation diagram, the inputs and outputs of a process are known. A process energy or mass balance is based on the fact that under steady-state operation, what goes into the process must come out. For example, when blending two flow streams using a mixer, the concentration of material dissolved or suspended in the liquid (expressed as the weight percent of the liquid steam) of the mixer outlet stream varies in a non-linear fashion with the inlet flow rates and inlet stream concentrations. The simulation diagram for a mixer process is shown in Figure 15-14.

The outlet stream concentration may be calculated based on a mass balance around the mixer, and blocks may be added to provide a dynamic

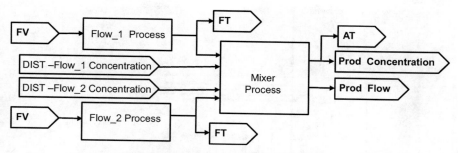

Figure 15-14. Simulation Diagram for a Mixer Process

response to input changes. The outlet concentration may be calculated as follows:

$$Solids_{in} = Solids_{out} = F1 \times C1 + F2 \times C2$$

$$Mass_{in} = Mass_{out} = F1 + F2$$

$$Outlet\ Concentration = \frac{Solids_{out}}{Mass_{out}} = \left(\frac{F1 \times C1 + F2 \times C2}{F1 + F2}\right)$$

where:

$F1$	=	inlet mass flow$_1$
$F2$	=	inlet mass flow$_2$
$C1$	=	solids concentration in flow$_1$
$C2$	=	solids concentration in flow$_2$

As this shows, the process gain (i.e., the change in mixer outlet concentration for a change in the flow rate of inlet flow to the mixer) depends on the flow rate and the concentration of both streams. This non-linear response can be included in the process simulation by calculating the outlet concentration based on the steady-state mass balance. Any transport delay or lag due to mixing may be accounted for using a deadtime and filter block in combination with this calculation as is illustrated in the mixer simulation composite shown in Figure 15-15.

When a combustion process (as described in Chapter 4) is to be simulated, it is often necessary to simulate the oxygen concentration in the exhaust gas. Over a wide operating range, the oxygen concentration in the flue gases, as measured by an O_2 analyzer, varies in a non-linear fashion with the fuel input rate and total air flow. This is another example of where the simulation range may be extended by accounting for this non-linear response in the simulation. The simulation diagram for a combustion process is illustrated in Figure 15-16.

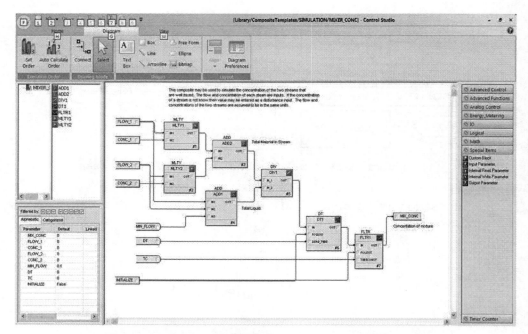

Figure 15-15. Mixer Outlet Concentration Simulation

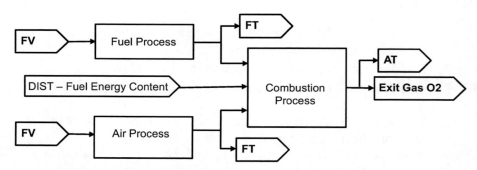

Figure 15-16. Combustion Process Simulation Diagram

To accurately simulate the flue gas O_2 concentration over a wide operating range, the calculation of O_2 using a mass balance may be combined with the lag and deadtime identified in the process step response. The oxygen concentration in air is approximately 21 percent by volume. Thus, the steady-state O_2 concentration after combustion may be calculated as follows.

$$Consumed\ Air = Fuel\ Flow \times K = Air\ Flow \times \left(1 - \frac{O_2\%}{21}\right)$$

where

$$K \quad = \quad \text{Air required per unit of fuel for complete combustion}$$

Solving for O_2, the following is obtained:

$$O_2\% = 21 \times \left(1 - \frac{K \times Fuel\ Flow}{Air\ Flow} \right)$$

As mentioned above, the impact of a change in fuel or air flow on the exit gas O_2 concentration varies with the fuel and air flow rates. To account for this non-linear response, the combustion process O_2 simulation may be based on the steady-state relationship between O_2, fuel flow and air flow. The calculated value must be limited to 0–21% to accommodate abnormal situations such as zero air flow. A deadtime and filter block may be used to account for transport delay and the mixing of gases in the combustion process as illustrated in Figure 15-17.

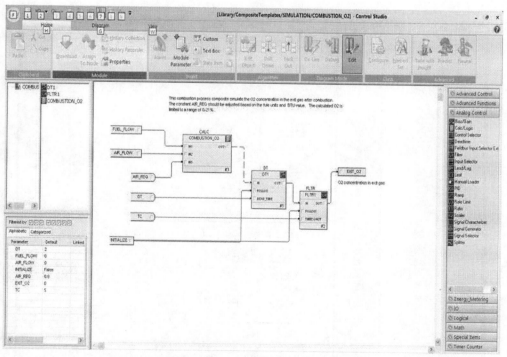

Figure 15-17. Combustion Process O₂ Simulation Composite

The simulations for the process shown in Chapter 16 demonstrate how non-linear process response may be accurately simulated by combining the results of steady-state analysis with the process step response.

15.4 Other Considerations

When using process simulation for control system checkout or operator training, it may be desirable to speed up the process response. A simulated process may be designed to provide faster than real-time response. This can be achieved by reducing the time constants and deadtime of a self-regulating process. Also, the integrating gain of an integrating process can be increased and the deadtime reduced. When the process response is modified in this manner, it is often necessary to re-tune the PID control to match the modified process response in the simulation.

Some control systems provide simulation applications that allow process simulation modules and control modules to be run faster than real-time, thus eliminating the need to modify simulation parameters and control tuning. Such tools may also allow the input block Simulate parameter to be enabled or disabled in all modules or in selected modules with one click of a button. An example of the interface provided for such a simulation environment is shown in Figure 15-18.

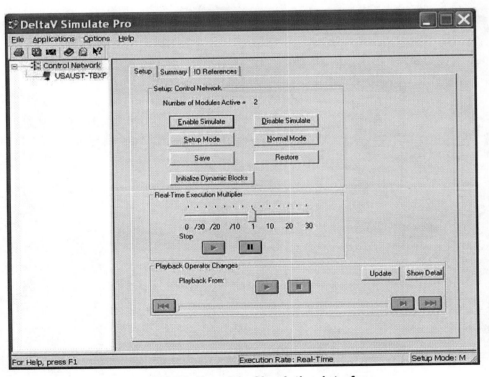

Figure 15-18. Example Simulation Interface

Many excellent commercial simulation products may be purchased for process design. The simulation tools supported by these products are based on first principle analysis of process equipment.

> **First Principle Analysis** – Analysis based on established laws of physics and chemistry that does not make assumptions such as empirical model and fitting parameters.

These process simulations typically provide better results over a wider operating range than is possible using a step response simulation model. Though such tools have been successfully used for operator training [1], the cost of engineering a simulation is often prohibitive if the simulation is to be used only for control system checkout or operator training. Also, the expertise needed to update these types of simulations to reflect changes made at plant startup is often not available in process plants.

15.5 Workshop – Process Simulation

Spray dryers are used in the process industry to manufacture a variety of products. The drying mechanism within the spray tower is quite complex and is impacted by the slurry flow rate, the spray pressure, the air flow, and the air temperature. This workshop exercise is designed to show how the spray dryer process may be easily simulated. See the Appendix for directions on using the web-based interface.

Step 1: Identify the small processes associated with the spray dryer process and circle these on the P&ID shown in Figure 15-19.

Step 2: Create a simulation diagram that consists of composite blocks that represent each process identified in Step 1.

Hint: The sprayer pressure is a function of the flow rate and how freely the feed flows through the spray nozzles. At a given flow rate, the pressure will increase as the nozzles start to plug. Thus, nozzle plugging can be considered to be a disturbance to the spray pressure process.

The control module and process simulation for the web exercise portion of the workshop is shown in Figures 15-20 and 15-21.

Step 3: In the Process Simulation workspace, change the slurry flow setpoint and observe how the heater outlet temperature is automatically adjusted to maintain the product moisture at setpoint.

Step 4: Place the heater temperature control in manual and then make a change in the slurry flow setpoint. Try to manually adjust temperature

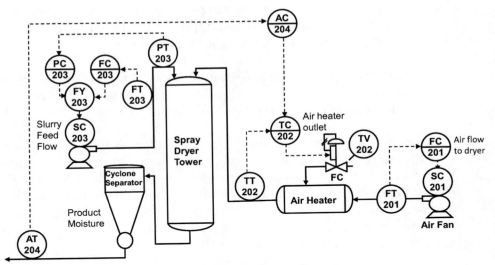

Figure 15-19. Spray Dryer Process

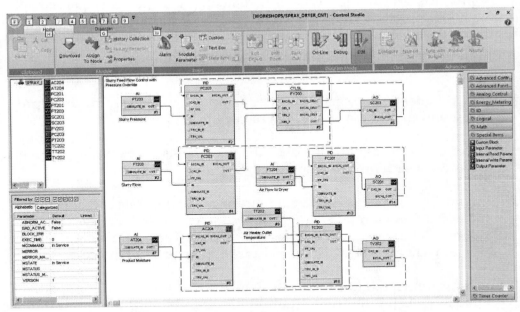

Figure 15-20. Spray Dryer Control Module

controller output to correct for this change in slurry flow. Place the heater temperature control base in Cascade mode and observe how the product moisture is brought back to setpoint.

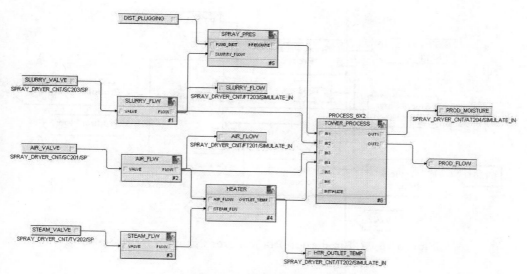

Figure 15-21. Spray Dryer Simulation

Workshop Discussion

The equipment associated with the spray dryer process can be logically broken into five processes that can each be modeled based on step response:

1. Slurry Flow Process
2. Spray Pressure Process
3. Air Flow Process
4. Air Heater Process
5. Tower Process

The simulation diagram based on this selection is shown in Figure 15-22. The slurry and air flow processes could be simulated in the same manner as the VALVE_FLOW composite. The air heater and spray pressure processes could be modeled in a similar way to the composite used to demonstrate feedforward, that is, a 2x1 process. The tower process could be simulated using a general step response model, that is, a 4x2 process that would be implemented in a similar way to the 2x2 process presented earlier in this chapter.

The spray pressure process is non-linear over a wide operating range. To illustrate this in the simulation, a characterizer block could be used to define the process gain as a function of the process flow rate.

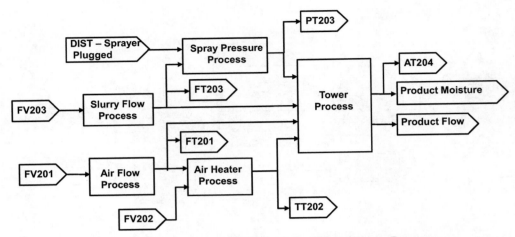

Figure 15-22. Simulation Diagram for Spray Dryer Process

References

1.   Blevins, T., McMillan, G. K., and Wheatley, R. "The Benefits of Combining High Fidelity Process and Control System Simulation," Proc. ISA EXPO 2000, Chicago, Oct. 2000.

16

Applications

The control systems commonly used in the process industry are based on the single-loop and multi-loop techniques described in previous chapters. Some combination of these techniques is often required to meet product specification and plant throughput targets. The application examples in this chapter are designed to illustrate the various ways these techniques may be used to automate plant control. The workshop exercise included at the end of each major section is based on one of the process examples. A dynamic process simulation is included with the process control module used in the exercise to allow you to get experience working with the control module and to explore the different features of the control design.

16.1 Inventory Control

Liquid storage tanks are used in various ways to control inventory within a plant. The feedstocks used in a plant may be shipped by rail car, barge, truck, or pipeline and stockpiled in storage tanks. The final product(s) produced by a plant may also be accumulated in storage tanks before being shipped to the customer. In addition, enough storage capacity of raw materials and products must be provided to allow manufacturing to continue between scheduled incoming and outgoing shipments.

Intermediate liquid storage tanks (surge tanks) are commonly installed between processing areas of the plant. Under normal operating conditions, these storage buffers allow each process area to be operated independently of the other process areas. Any off-spec or recycled material, or a material that is a by-product of manufacturing, may also be transferred to storage for further processing. In addition, within a process area there may be a requirement to maintain a specific quantity of liquid inventory for correct operation.

The inventory associated with feedstock and product tanks is most often maintained by an operator manually adjusting the target plant throughput. Similarly, the inventory in intermediate storage tanks between process areas may be maintained by an operator manually coordinating the throughput of each process area. However, improved coordination of process areas can often be achieved by automating the control of intermediate storage. Also, in most cases, automatic control is required to achieve specific liquid inventory levels within a process area. In this section, some examples are shown where inventory control has been automated.

16.1.1 Surge Tank

Any imbalance in the area production rates within a plant will be reflected by a change in level of the surge tanks between process areas. When a downstream process area is made up of a continuous process, then its throughput may be automatically adjusted based on the surge tank level, as illustrated in Figure 16-1.

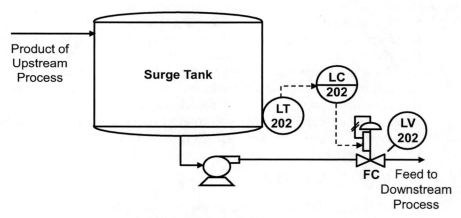

Figure 16-1. Surge Tank Level

To avoid abrupt changes in production rate in the downstream process, the level controller (LC 202) may be configured for proportional-only control, with the bias setting based on the normal plant production rate. The range over which this floating-level control adjusts the downstream flow is determined by the controller gain and bias. Alternatively, PI control may be used where the reset determines the time required for any change in the upstream process throughput to be reflected in the downstream process throughput.

16.1.2 Recycle Tank

As the result of a manufacturing process, a by-product may be created that in turn can be used as a feedstock within the process. For example, the steam condensate from an evaporator can be used in a process area that requires a water supply. In such cases, the by-product is typically transferred to a recycle tank. To account for any imbalance in the recycle supply and the process feed requirements, the recycle tank makeup stream may be automatically regulated as illustrated in Figure 16-2.

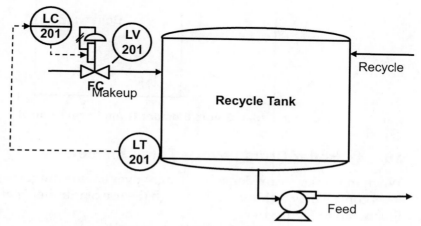

Figure 16-2. Recycle Tank Level

In this case, the level control could be configured as a proportional-only control where the bias setting is based on the normal makeup flow rate. This floating-level control ensures that makeup is added only if the recycle tank level drops below a point determined by the bias and the controller gain. When level control is configured in this manner, little or no operator input is required during normal operation.

16.1.3 Boiler Drum Level – Single Element

The correct operation of a boiler depends on careful regulation of the level of water in the boiler drum. If the water level is allowed to drop too low, then this may disrupt the flow of water to the steam generating tubes and potentially cause tube failure. If the water level is allowed to rise too high, then the resulting water carryover into the steam leaving the boiler can harm downstream turbines or other users of steam.

When the steam demand is fairly constant, it is possible to maintain boiler drum level using single-loop control as illustrated in Figure 16-3. Such an approach is often referred to as single element drum level control since only one measurement is used in the control implementation.

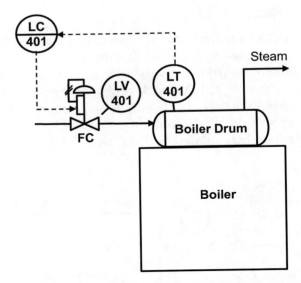

Figure 16-3. Single Element Drum Level Control

16.1.4   Boiler Drum Level – Three Element

When the boiler steam demand changes over a wide range, boiler drum level control may be improved through the use of multi-element control as illustrated in Figure 16-4.

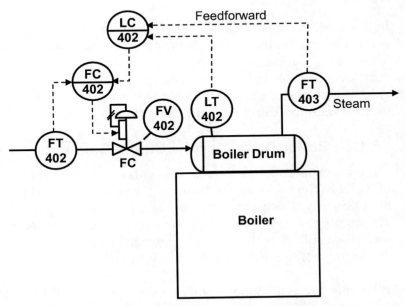

Figure 16-4. Three Element Drum Level Control

Changes in the boiler steam flow resulting from variations in demand are a disturbance to the level control process. Thus, variations in the drum level may be reduced by introducing the steam flow measurement as a feedforward input to the level control. Also, the impact of any non-linear installed characteristic of the feedwater valve on the level control may be minimized by measuring the feedwater flow rate and structuring the level control as a level-flow cascade loop. Power boilers often use three element drum level control. Also, three element drum level control is standard on most large utility and waste fuel boilers.

16.1.5 Workshop – Three Element Drum Level Control

Explore the Three Element Drum Level Control workshop based on Figure 16-5. See the Appendix for directions on using the web-based interface.

Step 1: Using the web-based interface, open the Three Element Drum Level Control workshop.

Step 2: In the workspace shown in Figure 16-5, observe the response to changes in setpoint and steam flow. Does the change in steam flow impact the drum level?

Step 3: Place the level control in manual mode and observe the impact on the drum level as steam flow changes.

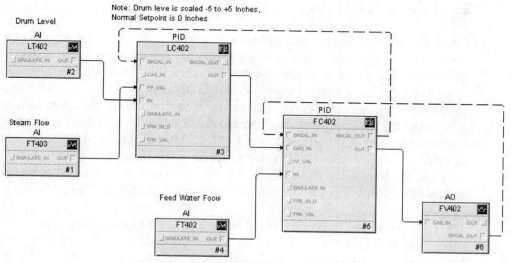

Figure 16-5. Inventory Control Workshop Module

16.2 Batch Processes

Batch processing is used in all sectors of the process industry, with small, local batch processes often being found in continuous process plants. One major advantage that batch processing has over continuous processing is that a wide variety of products can be manufactured in small quantities with the same equipment. Also, when a process's chemical reaction or biological growth rate is slow, then batch processing may be the only practical means of achieving the required retention time needed to manufacture a product. Thus, batch processing is frequently used in specialty chemicals and pharmaceutical manufacturing.

The control loops applied to batch processes are based on both single-loop and multi-loop control. Some examples are provided in this section to illustrate how these strategies may be used in the control of batch processes.

16.2.1 Batch Digester

Batch digesters are vessels used in the manufacture of paper products. At the start of a batch cycle, the capping valve (a large on-off valve at the top of the digester) is opened and wood chips are fed into the digester using a belt conveyor. When the chips inside the digester reach a target level, the conveyor belt is stopped, the capping valve is closed and a specific volume of a chemical known as white liquor is added to the digester. White liquor primarily consists of sodium hydroxide. This strongly alkaline (caustic) chemical is used to break down and remove the lignin that binds cellulose fibers within wood chips.

Once the digester is charged (filled) with chips and white liquor, the contents of the digester are raised to an elevated temperature to speed the delignification process. A temperature control loop that adjusts the steam flow to an external heater is used to automatically bring the digester temperature to a target value. Any gases and non-condensable vapors that are generated by the chemical reaction are automatically removed by a pressure control loop that adjusts a vent valve. After a period of time, the contents of the digester are removed by opening the blow valve—an on-off valve at the bottom of the digester vessel. The control loops and measurements commonly found on a batch digester are shown in Figure 16-6.

The chip conveyor and equipment for white liquor charge are often common to all the digesters in the plant. Thus, the charging of digesters must be coordinated by batch control software. Also, the pulp produced by the digester is often transferred to a common blow tank and this transfer must be scheduled by the batch control software.

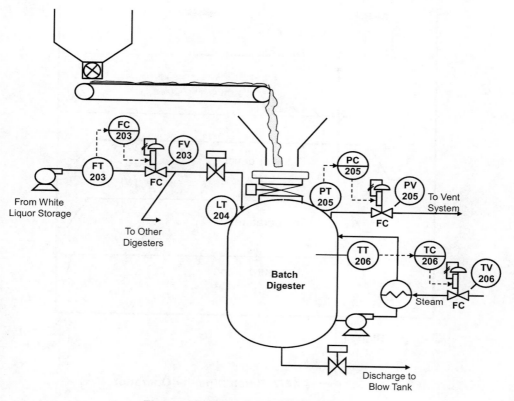

Figure 16-6. Batch Digester Control

In discussing the processing done by the batch control software, it is helpful to document the desired sequencing of events and changes to process variables during the batch operation. The processing associated with the batch digester can be diagrammed as illustrated in Figure 16-7.

16.2.2 Batch Chemical Reactor

Batch chemical reactors may be used to manufacture a variety of products. Typically the reactor is initially charged (filled) with one or more chemicals, the contents of the reactor are heated to a target temperature and then a metered quantity of one or more reactive chemicals is added to the batch. The chemical reaction may be exothermic and thus cooling may be required to maintain the reactor charge temperature at the target temperature. A different reaction may be endothermic, and thus heating will be required. In order to support both heating and cooling, temperature control is implemented as a split-range control loop that adjusts heating and cooling media flow to the reactor shell. Similar to the batch digester, the reactor pressure is maintained at setpoint by venting vapors and gases generated by the chemical reaction.

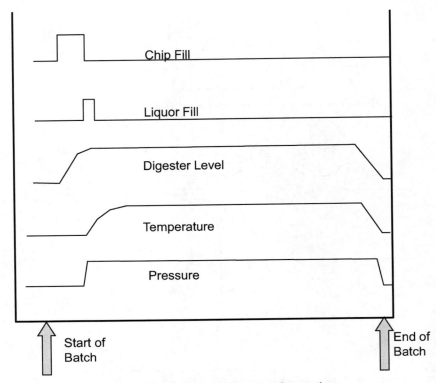

Figure 16-7. Batch Digester Operation

With an exothermic reaction, when the flow rate of the reactive chemical(s) to the reactor increases, more cooling is required. If the cooling requirements exceed what can be provided by the cooling media with the cooling valve wide open, then the batch temperature will exceed the temperature controller setpoint—TC211 in Figure 16-8. If the temperature continues to rise, when the temperature reaches the setpoint of the override controller (TC211A) the feed flow rate to the reactor will be reduced to prevent the temperature from exceeding the override temperature setpoint. If the production of gas and vapor exceeds the capacity of the vent system, then when the pressure reaches the setpoint of the override controller (PC212) the feed flow rate to the reactor will be reduced to prevent the pressure from exceeding the override pressure setpoint. The control loops and measurements commonly found on a batch reactor are shown in Figure 16-8.

The feedstock tank and valve manifold used to initially charge the reactor are often shared among multiple reactors. Thus, the timing of the reactor charge must be coordinated by the batch control software. The sequence of such a batch process may be illustrated as shown in Figure 16-9.

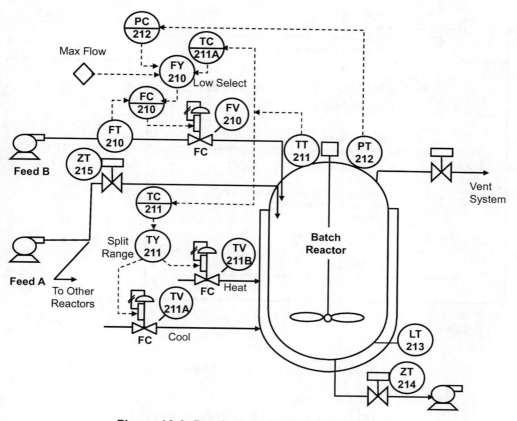

Figure 16-8. Batch Chemical Reactor Control

The feed rate of reactive chemical(s) to a batch chemical reactor is often set by the operator to a value that ensures that the capability of the cooling and vent systems will not be exceeded (i.e., the pressure and temperature overrides will remain inactive throughout the batch). Thus, at times a higher feed and throughput rate could be achieved since the capacity of the cooling and vent systems which can support multiple reactors may vary with plant operating conditions, for example, cooling media temperature. Using Model Predictive Control (MPC), it is possible to operate at the temperature and pressure operating constraints to achieve the maximum possible feed rate and reduce the batch cycle time as illustrated in Figure 16-9. The use of MPC is shown in Figure 16-10.

As previously discussed, MPC could be layered on top of a traditional control system.

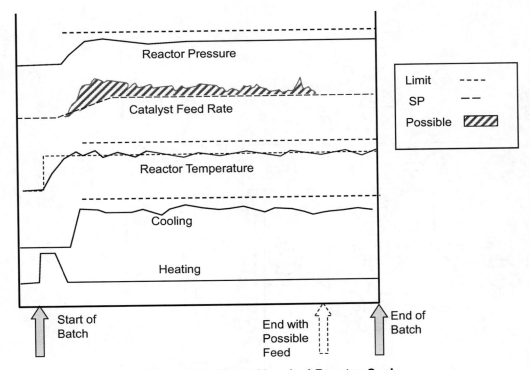

Figure 16-9. Batch Chemical Reactor Cycle

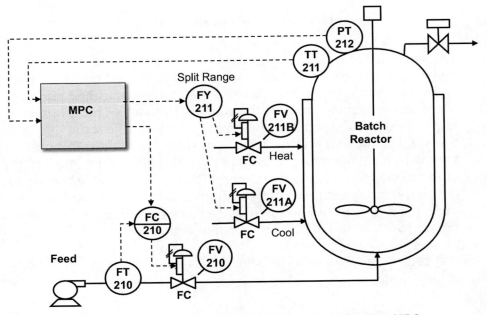

Figure 16-10. Batch Chemical Reactor Control Using MPC

16.2.3 Batch Bioreactor

The pharmaceutical industry manufactures a wide variety of drugs using batch processes. A product area of growing importance is one in which a mammalian cell culture is grown in a bioreactor. The measurements and controls for such a bioreactor [1] are shown in Figure 16-11.

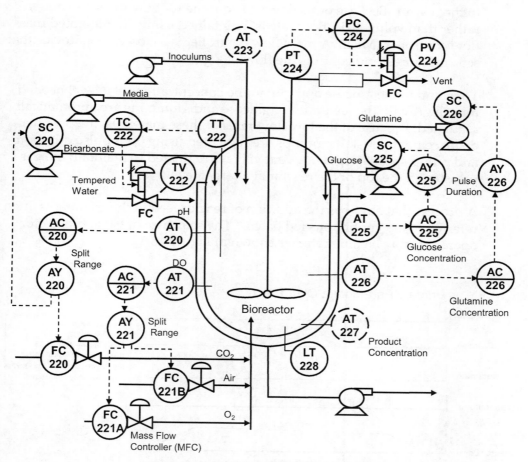

Figure 16-11. Batch Bioreactor Control

During the start of the batch cycle, media is added to a sterile reactor until a target volume is reached, as indicated by the reactor level.

> **Media** - A liquid or gel designed to support the growth of microorganisms or cells.

An initial charge of cells, known as the inoculums, is then added to the reactor. During the batch cycle, the reactor temperature is tightly maintained by regulating a tempered water flow through the jacket of the reactor. Also, the pH needed for cell growth is automatically maintained through the split-range adjustment of CO_2 flow rate and bicarbonate of soda feed pump speed to the reactor.

As the cells multiply within the reactor, the required dissolved oxygen content must be maintained through the split-range addition of air and O_2 to the reactor. Because of the low CO_2, air and O_2 flow rates, the mass, rather than volumetric, flow rate is maintained using an integrated mass flow controller (MFC). A mass flow controller is a closed-loop device that sets, measures, and controls the flow of a gas or liquid.

To maintain the glucose and glutamine concentrations at a level needed for cell growth, the feed rates of glucose and glutamine are automatically regulated. Because of the low flow rates of these materials, the positive displacement peristaltic pumps are regulated using both pulse duration and pump speed outputs. A vent valve is regulated to maintain the gas pressure in the top of the reactor at a target level.

A process that employs the addition of reactor feed in this manner is often referred to as "continuous fed batch." The sequence of batch bioreactor operation may be illustrated as shown in Figure 16-12.

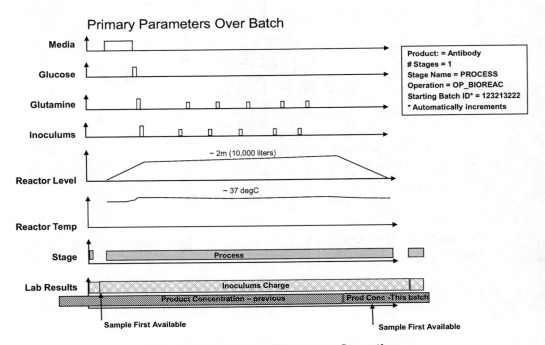

Figure 16-12. Batch Bioreactor Operation

In some cases, the measurement of glucose and glutamine level may be available only through off-line lab analysis. If this is the case, then the duty-cycle pulse duration of the glucose and glutamine feed pumps is manually adjusted by the operator to maintain a target flow rate based on the latest lab analysis.

The end point of a bioreactor batch may be established based on the batch processing time or on the cell concentration level as measured by the lab or an on-line analyzer. The contents of the batch are removed from the reactor using a peristaltic pump.

16.2.4   Workshop – Batch Chemical Reactor

Explore the Batch Chemical Reactor workshop, based on Figure 16-8. See the Appendix for directions on using the web-based interface.

Step 1: Using the web-based interface, open the Batch Control workspace shown in Figure 16-13 and observe the changes in batch ID and control setpoints made by the batch control. Examine the chart to see when targets are changed in the batch.

Step 2: Increase the maximum feed rate target, MAX_FLOW, and observe the impact on the batch operation.

16.3 Continuous Processes

16.3.1   Chemical Reactor

A continuous chemical reactor process such as that shown in Figure 16-14 is characterized by the continuous feed of two or more chemical reactants and a continuous output flow of product. To achieve the desired product specifications, the proportions of each reactant must be tightly maintained at a target value using ratio control.

The *chemical reaction* that occurs within the reactor is often exothermic. In such cases, the reactor temperature may be maintained at setpoint by automatically regulating coolant flow through an external heat exchanger. Any gases and vapors that are created as a by-product of the chemical reaction are vented to maintain the reactor overhead pressure at setpoint.

Any imbalance between the feed stream input flows and the product flow from the reactor is reflected by changes in the reactor level measurement. Thus, the product flow rate is automatically regulated to maintain the reactor level at setpoint.

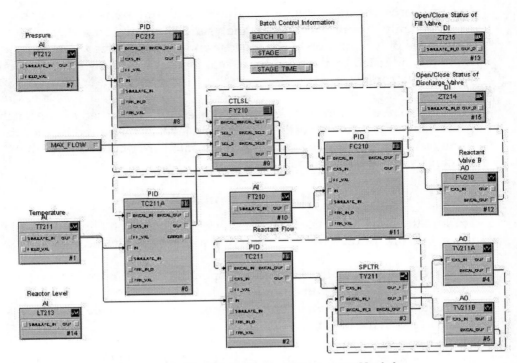

Figure 16-13. Batch Control Workshop Module

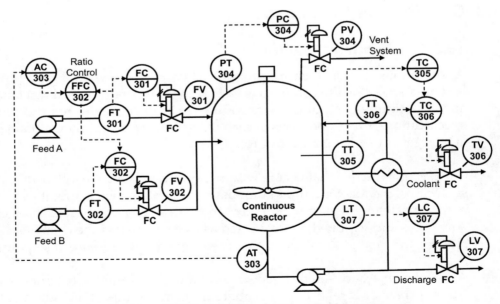

Figure 16-14. Continuous Chemical Reactor

Any changes in the composition of the feedstock streams will impact the rate of chemical reaction and the composition of the product stream. To compensate for this type of change, the feed stream ratio may be automatically regulated based on an on-line analysis of the product. If the measurement provided by this analysis deviates from setpoint, then the analytic controller will adjust the feed stream ratio to bring the product composition measurement back to setpoint. If the product is analyzed off-line in a lab, then the lab results may be used by the operator to manually adjust the feed stream ratio setpoint.

16.3.2   Spray Dryer Control

Spray dryers are used in the chemical, pharmaceutical, food, and dairy industries to manufacture a wide variety of products. In a spray dryer process, a slurry feed is sprayed into the dryer and the liquid component is removed by a hot gas stream to create a powdered product. The dryer design and the type of spray nozzles (or atomizers) are selected based on the properties of the slurry to be dried and the characteristics of the powder.

The slurry feed rate is normally set by the operator. However, if the spray nozzles begin to plug, then the pressure to the nozzles will increase. Under this condition, the feed flow rate may be automatically reduced using pressure override control to help maintain consistent particle size in the spray. Any reduction in flow rate will impact the air temperature required to dry the material.

The air flow rate to the dryer is set based on the target slurry feed flow rate and operating conditions. The temperature of the air to the dryer is automatically controlled by regulating the air heater fuel input.

Any significant change in the feedstock concentration, feed rate or air flow to the dryer will impact the residual moisture content of the product. If the product residual moisture content is measured on-line, then moisture control may be cascaded to the air heater temperature loop as illustrated in Figure 16-15. The air temperature is automatically adjusted to bring the moisture content back to setpoint.

When the moisture content is only available from the lab, the lab measurement may be used by the operator to manually change the air heater temperature setpoint.

As indicated, any change in the dryer feed flow, sprayer pressure, air temperature, or air flow will impact the product residual moisture. In some cases, Model Predictive Control has been successfully used to automati-

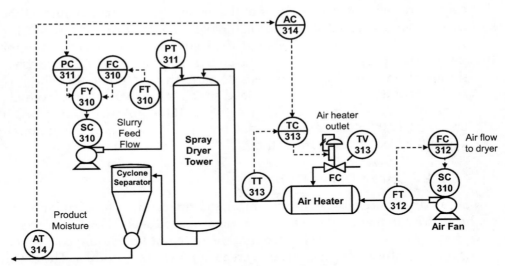

Figure 16-15. Spray Dryer

cally account for the impact of these changes on product residual moisture as shown in Figure 16-16.

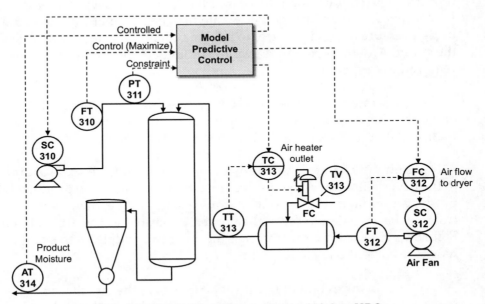

Figure 16-16. Spray Dryer Control Using MPC

Through the use of MPC, it is possible to maximize dryer throughput by maintaining the feed rate at a value that maintains the dryer at its spray pressure constraint. Also, any delays in a change in process inputs being reflected in product moisture content are automatically compensated for when MPC is used to automate the spray dryer.

16.3.3   Workshop – Continuous Chemical Reactor

Explore the Continuous Chemical Reactor workshop based on Figure 16-14. See the Appendix for directions on using the web-based interface.

Step 1: Using the web-based interface, open the Continuous Control workshop.

Step 2: In the workspace area shown in Figure 16-17, change the Feed setpoint (FC301) and observe the response.

Step 3: Change the composition (AC303) setpoint and observe the response.

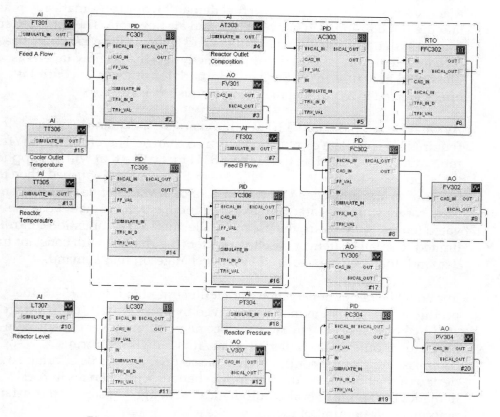

Figure 16-17. Continuous Control Workshop Module

16.4 Combustion Control

One of the first areas in the process industry to use automated control was combustion control in the powerhouse area. To meet changing process steam demands, it was—and is—necessary to automate the control of power boilers. Today small boilers, heaters, and power boilers are all automatically controlled.

The design of the control system for a combustion process has a large impact on boiler and heater efficiency and operating safety. In this section, the combustion control systems of a variety of processes are shown.

16.4.1   Small Boiler/Heater

For safe and efficient operation and to meet environmental regulations, the air supply to any combustion process must be sufficient to burn all the fuel input. Some of the early combustion control systems were based on the fuel valve and air damper linkages being mechanically linked through a jack shaft, so that any change in fuel demand would automatically maintain the same proportion between fuel and air flow rate. The combustion control systems of some smaller boilers and heaters in use today are based on equivalent ratio control (i.e., fuel directly sets the air flow target) as illustrated in Figure 16-18. Such an approach to combustion control is often referred to as parallel metered combustion control.

As we have seen, if the fuel BTU value changes, the quantity of air required for complete combustion will also change. Therefore, the air/fuel ratio must be set based on the highest BTU fuel value that will be supplied to the boiler or heater. When fuel of lower BTU content is burned, then the boiler or heater efficiency will decrease since any air not needed for complete combustion is heated and exits the combustion chamber at an elevated temperature. Thus, the use of parallel metered combustion control is limited to small heaters or base load boilers (i.e., boilers with constant fuel demand) that are characterized by slow changes in fuel demand.

In a small heater or boiler, such as that illustrated in Figure 16-18, the products of combustion may be removed by natural draft (i.e., draft created by hot gases rising in the stack because of their lower density compared to air outside the boiler). The air/fuel ratio is set during commissioning of the control system based on the fuel's BTU value and design of the burner. As the fuel input changes with changes in fuel demand, the air input is also automatically changed to supply a constant amount of air per unit of fuel.

One significant disadvantage of a parallel metered system is that it may be possible to increase or decrease fuel input faster than the air flow can

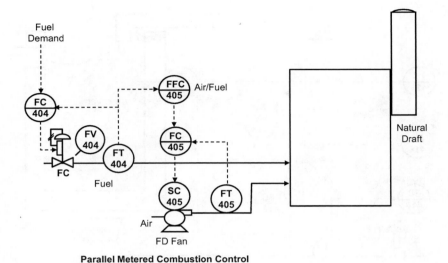

Parallel Metered Combustion Control

Figure 16-18. Small Boiler/Heater Combustion Control

respond which allows more fuel to be added without sufficient air to support combustion of all the fuel. Thus, to avoid an explosive, fuel rich condition with a rapid increase in fuel input rate it may be necessary to limit the rate at which the fuel input may be increased. Interlocks must also be provided to ensure that the fuel flow is cut off when air flow is lost.

16.4.2   Vat Heater

A gas or oil fired heater may be used to heat a liquid that is used in a manufacturing process. To account for the differences in the speed of response of the fuel and air flows to changes in manipulated inputs, a lead-lag combustion control system may be used. This type of combustion control is illustrated in Figure 16-19 for a vat heater application. In this case, the fuel demand is regulated by a temperature controller to maintain vat temperature. At the liquid flow rate to the vat changes then the temperature controller will change the fuel demand to maintain target temperature.

The heart of the lead/lag combustion control system is the air/fuel ratio controller. However, additional components are added to ensure that on an increase in fuel demand and the resulting increase in fuel flow, the change in air flow will always lead the fuel flow and that on a decrease in fuel demand, the change in air flow will always lag the change in fuel flow. To achieve this behavior, the air supply setpoint is calculated based on the air/fuel ratio and the higher value of the fuel demand and the fuel flow measurement. Similarly, the fuel controller setpoint is calculated based on the lower value of the fuel demand and the fuel flow that can be supported by the air flow to the burner.

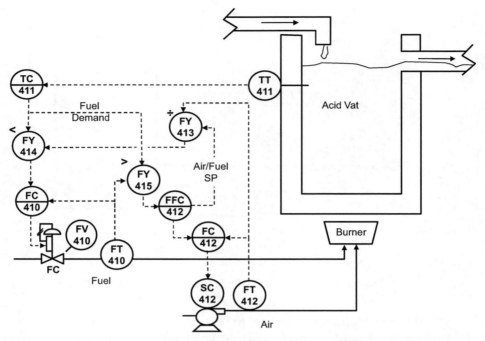

Figure 16-19. Vat Heater

$$Fuel\ flow\ that\ can\ be\ supported\ = \left(\frac{Air\ flow\ measurement}{\dfrac{Air}{fuel}Ratio\ SP}\right)$$

Another benefit of the lead-lag combustion control system is that the fuel flow setpoint will be automatically set to zero on the loss air flow, independently of the fuel demand.

16.4.3   Power Boiler – Single Fuel

A power boiler is designed to operate over a wide range of steam demand and to respond to rapid changes in fuel demand. One or more power boilers may be used to meet the changing steam demand of the plant. A lead-lag combustion control system that might be found on a power boiler that burns a single fuel is illustrated in Figure 16-20.

The fuel demand from the plant master (see powerhouse example in Chapter 13) passes to the boiler through a bias/gain station, known as the boiler master. Using the boiler master, the operator may bias the plant master fuel demand to determine the boiler fuel demand. Such operator intervention may be needed if there is a condition that limits the boiler steam production or if it is more economical to generate steam in one

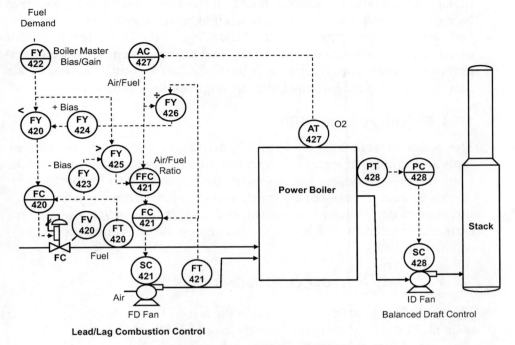

Figure 16-20. Power Boiler – Single Fuel

boiler versus another boiler. Also, by placing the boiler master in Manual, it is possible for the operator to manually set the boiler fuel demand. At a constant fuel demand, the fuel and air flow controllers will be at setpoint.

Two BIAS blocks are included in this lead-lag combustion control system that were not shown in the vat heater example. The bias values are used to prevent a value based on air or fuel flow from being selected because of measurement noise. The bias value is set equal to the noise level in the associated measurement.

As shown in Figure 16-20, an induced draft (ID) fan may be required to remove the products of combustion from a larger boiler. A negative pressure is maintained in the boiler to prevent hot gases from being blown out through access ports and openings in the boiler. The boiler draft may be maintained at a negative pressure by regulating the speed of the ID fan.

As was discussed in Chapter 4, air contains approximately 21% oxygen by volume, with the remaining components primarily being nitrogen. To ensure complete combustion, some excess air must be supplied to the combustion process to provide proper mixing. As the fuel BTU content changes, the quantity air required for complete combustion will change. To compensate for BTU content changes and allow the boiler to achieve maximum operating efficiency, the O_2 in the gases leaving the boiler may be measured and the measurement used to automatically adjust the air/fuel ratio. It is good practice to limit the range of the O_2 controller output to a safe minimum and maximum air/fuel ratio. This ensures safe operation even if the O_2 analyzer fails or the O_2 measurement is skewed by air that has been pulled into the boiler through an open access door.

16.4.4 Rotary Lime Kiln

The design of a combustion control system will vary with the number of fuels and sources of air and the process measurements that are available. Also, the process design and operating requirements can impact the control system design. For example, a combustion process is used in a lime kiln to dry a slurry feed stream and to heat and convert the calcium carbonate content to quicklime, calcium oxide, through the process of calcination:

$$CaCO_2 \Rightarrow CaO + CO_2$$

The transfer of heat to the material in the hot end of the kiln (Figure 16-21) is by radiant energy transfer from the combustion gases.

Gas or oil may be burned in the kiln. A forced draft fan is used to provide some of the air required for combustion and to shape the flame within the kiln. However, most of the air for combustion is normally pulled into the kiln by the ID fan and is pre-heated as it travels over the hot material leaving the kiln. The ID fan also must maintain a negative draft in the kiln to avoid hot gases blowing back through access ports in the firing end of the kiln. A two-color pyrometer is used to measure the material temperature within the lower end of the kiln. The exit gas temperature and oxygen content are measured as illustrated in Figure 16-21.

To maintain the correct flame shape as fuel input changes to maintain kiln temperatures with varying slurry feed rate, the FD (forced draft) fan speed may be characterized as a function of fuel input using a characterizer block, FY430.

For best conversion, the temperature of the material leaving the kiln must be maintained above that required for the calcination process. For example, the temperature must be maintained above the dissociation temperature of the carbonates in the lime mud which is between 780°C and 1340°C (2, 3). Any imbalance in the amount of energy supplied by the fuel and that required for drying and to support calcination is reflected in the temperature of the gas leaving the kiln. The fuel input is regulated to maintain the exit gas temperature at the value that is required to dry and process the slurry feed. Since the total air flow the kiln is not measured, the ID fan

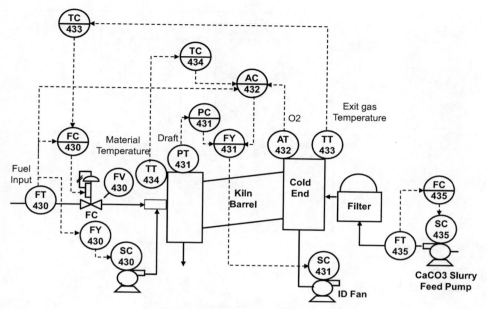

Figure 16-21. Rotary Lime Kiln

is regulated to maintain a target O_2 level in the gas in leaving the kiln. To ensure that a negative pressure is always maintained in the kiln, actions by the O_2 controller may be overridden by the pressure controller.

Since the flame temperature is a function of the excess air level, the material temperature is maintained at setpoint by regulating the O_2 setpoint over a narrow range that ensures complete combustion. Because of the interactions and operating constraints associated with the lime kiln process, it is often possible to increase throughput and improve product quality and operating efficiency through the use of MPC. [4]

16.4.5 Workshop – Power Boiler Control

Explore the Power Boiler Control workshop based on Figure 16-20. See the Appendix for directions on using the web-based interface.

Step 1: Using the web-based interface, open the Power Boiler workshop. In the workspace shown in Figure 16-22, observe the control response for a step change increase in fuel demand.

Step 2: Observe the control response for a step change decrease in fuel demand.

Step 3: Change the O_2 setpoint and observe the control response.

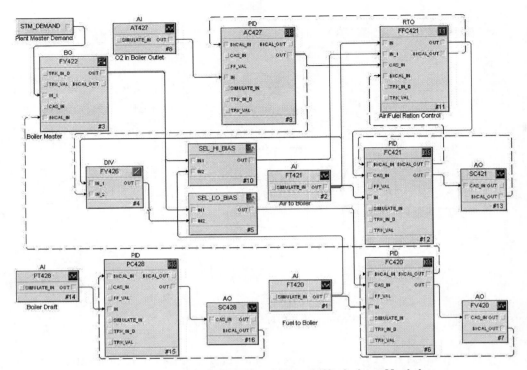

Figure 16-22. Combustion Control Workshop Module

16.5 Distillation Control

Continuous distillation is used in many industries to separate mixtures of liquids based on differences in their volatility. Within a distillation column, the vapor rising from the bottom of the column is used to heat the feed entering the column and the reflux material. Heat transfer between the vapor and liquid is increased by metal trays or by packing that fills the column. Liquid that falls to the bottom of the column is reheated to boiling using an external heater. The lighter material that reaches the top of the column as a vapor is condensed using a heat exchanger and collected in the accumulator. A portion of the condensate is removed from the accumulator as overhead distillate and the remaining liquid is returned to the column as reflux. Figure 16-23 shows one example of the controls that may be provided on a distillation column.

The extent to which the lighter components may be separated from the heavier component of the feed to a distillation column is determined by the column design, the properties of the feedstock and the energy input. By implementing a steam/feed ratio control the energy input per unit of feed may be maintained constant. The differences in the time required for a feed flow change and a change in energy input to impact vapor composi-

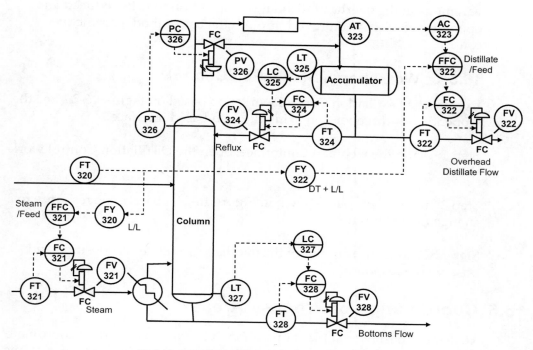

Figure 16-23. Distillation Column Control

tion before the accumulator can be addressed through a lead/lag block inserted in the feed flow input to the steam/feed ratio control.

As the feed rate increases, the overhead distillate flow should also be increased proportionately to achieve the same distillate composition. This balance is achieved using a distillate/feed ratio control. To compensate for changes in the feedstock composition, the distillate/feed ratio may be automatically adjusted based on an analysis of the distillate.

A balance between the condensed vapor flow to the accumulator and flow from the accumulator is maintained by regulating the reflux flow based on accumulator level. Similarly, a balance between the feedstock flow and the material removed from the column as overhead distillate flow is maintained by regulating the removal of liquid from the bottom of the column, through the bottoms flow, to maintain a constant level in the bottom of the column. Any imbalance between the rate of vapor generation in the column and the rate of vapor condensation in the heat exchanger is reflected in the column pressure. Thus, the vapor flow that bypasses the heat exchanger is regulated to maintain a constant column pressure.

The complex dynamics and interactions associated with a distillation column may make it difficult to commission and operate. In some cases, variations in the overhead distillate composition may be reduced and operating efficiency improved through the use of model predictive control.

16.5.1   Workshop – Distillation Control

Explore the Distillation Control workshop based on Figure 16-23. See the Appendix for directions on using the web-based interface.

Step 1: Using the web-based interface, open the Distillation Control workshop.

Step 2: In the workspace shown in Figure 16-24, observe the response to a change in feed input.

Step 3: Change the setpoint of the overhead distillate composition and observe the response.

16.6   Coordination of Process Areas

To achieve best overall plant operation, it is often necessary to coordinate the operation of equipment in multiple process areas. In this section, some examples are given where multi-loop techniques may be effectively used for such coordination.

16.6.1   Ammonia Plant H/N Control

The efficient operation of an ammonia plant (described in Chapter 10) depends on the hydrogen-to-nitrogen ratio in the synthesis loop at the back end of the process being tightly maintained. However, this ratio is determined by the gas and air feed to the primary and secondary reformers at the front end of the plant. Thus, the controls needed to maintain the H/N ratio require coordination between these plant areas, as illustrated in Figure 16-25.

The throughput of the plant is set by the rate of natural gas flow—primarily methane, CH_4—to the primary reformer. Steam must also be added in a fixed proportion to the gas flow:

$$CH_4 + H_2O \Rightarrow 2H_2 + CO$$

A steam/carbon ratio controller is used to regulate the steam flow based on the gas flow to the primary reformer. Air added to the secondary reformer is the source of nitrogen used in ammonia production, with the air flow requirement based on the gas flow to the primary reformer. The ratio between air flow and gas flow is maintained using an air/gas ratio

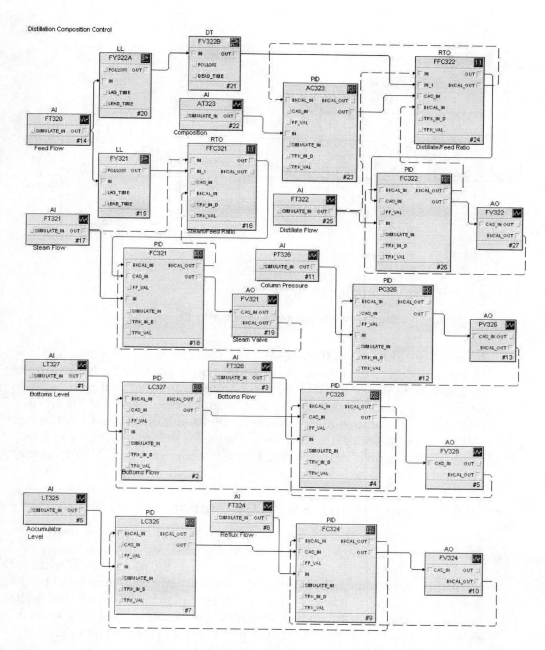

Figure 16-24. Distillation Control Workshop Module

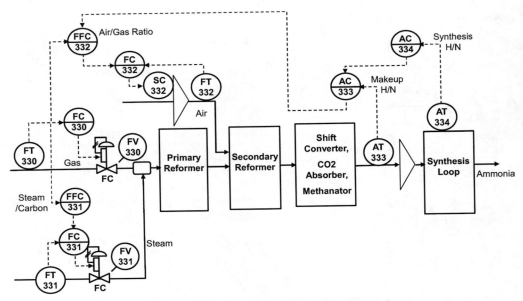

Figure 16-25. Ammonia Plant H/N Control

control. Any changes in the operation of the secondary reformer, and equipment before the synthesis converter will impact the H/N ratio in the feed stream to the synthesis loop. Thus, a measurement of the H/N ratio before the synthesis loop may be used to automatically adjust the air/gas ratio. The H/N ratio in the synthesis loop is maintained at setpoint by regulating the setpoint of the H/N ratio controller before the synthesis loop. Through this control, it is possible to maintain a precise H/N ratio in the synthesis loop.

16.6.2   Power House Steam Generation

As described in Chapter 13, one or more power boilers may be used to meet the process steam requirements of a plant. Also, this steam may be used in turbine generators to meet some or all of the plant requirement for electricity. Any imbalance between the quantity of steam generated by the power boilers and that required to operate the plant will be reflected in the pressure in the main steam header. Thus, a pressure control loop known as the plant master is used to set the power boiler steam demand as illustrated in Figure 16-26.

The high pressure steam generated by the power boilers is used to feed the turbine generators. The pressure of the steam that is extracted from the turbines at various points is automatically regulated by the turbine control system to meet the lower pressure steam requirements of the process. If a turbine goes off-line, then the pressure in the lower pressure steam headers will drop. Once the pressure drops to the setpoint of a header pressure

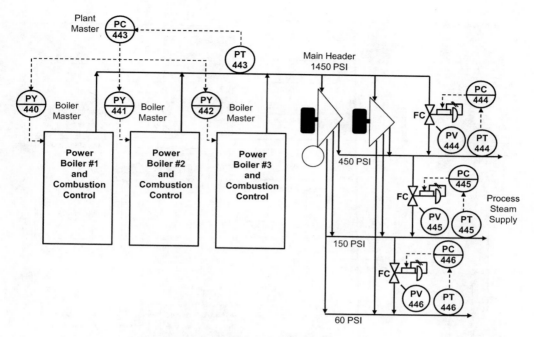

Figure 16-26. Power House – Steam Generation

controller, then a pressure reducing valve is opened to maintain the pressure at setpoint.

16.6.3   Workshop –Ammonia Plant H/N Control

Explore the Ammonia Plant H/N Control workshop based on Figure 16-25. See the Appendix for directions on using the web-based interface.

Step 1: Using the web-based interface, open the Ammonia Plant H/N workshop.

Step 2: In the workspace shown in Figure 16-27, observe the response for a step change increase in plant feed rate.

Step 3: Change the loop H/N ratio setpoint and observe the response.

16.7 Difficult Dynamics, Process Interaction

Single-loop and multi-loop control may be used to address a wide variety of control requirements. However, if a process is characterized by extremely long process delays or by a high degree of process interaction, then other control techniques may be required. In this section, two examples are provided that illustrate processes that are better addressed using other control techniques.

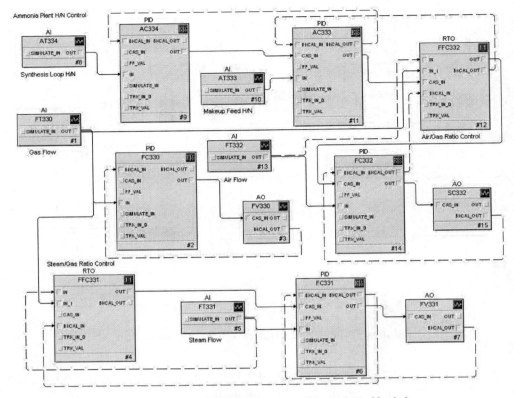

Figure 16-27. Unit Coordination Workshop Module

16.7.1 Pulp Bleaching

In the bleaching area of a paper pulp mill, as illustrated in Figure 16-28, bleaching chemicals are used to achieve a desired level of pulp brightness. Since the reaction after chemical addition is very slow, the bleach tower is designed for long residence times. Because of the transport delay associated with the bleach tower, a brightness measurement made after the bleach tower may not reflect a change in bleaching chemical addition for 30–60 minutes depending on the pulp feedstock flow rate.

To account for changes in the feedstock consistency, that is, changes in solids ("dry stock") concentration, and feedstock flow rate, the bleaching chemical addition rate may be set by ratio control of chemical/dry stock. However, to effectively adjust this ratio based on deviations of brightness from setpoint, it is necessary to use a control technique such as MPC that accounts for the very long process delay.

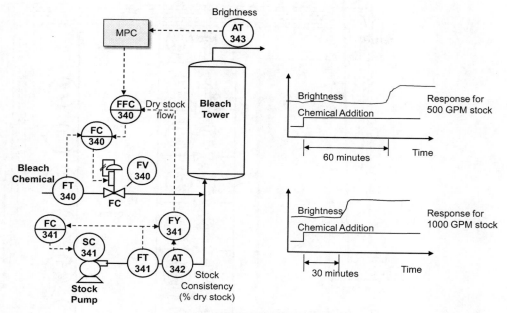

Figure 16-28. Pulp Bleaching

16.7.2 Primary Reformer Temperature

In an ammonia plant, the primary reformer is made up of multiple tube bundles, known as harps, which contain a catalyst. Gas burners are located between each harp at the bottom of the reformer. To drive the reforming process to completion, it is important to maintain a high gas temperature leaving the harp.

The outlet temperature of each harp is measured. However, any adjustment of one burner will impact the outlet temperature of adjacent harps. Such interactions are best addressed using MPC. MPC implementation is illustrated in Figure 16-29.

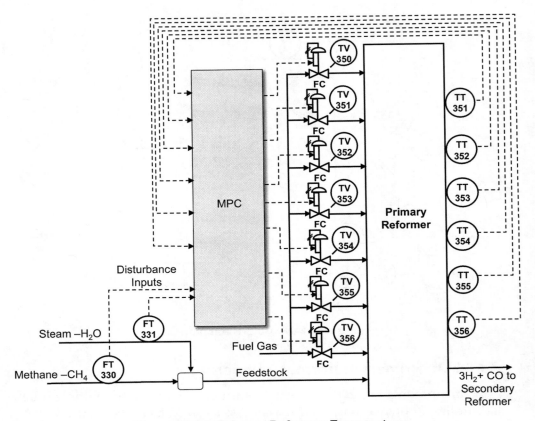

Figure 16-29. Primary Reformer Temperature

References

1. Boudreau, M. and McMillan, G. K., *New Directions in Bioprocess Modeling and Control: Maximizing Process Analytical Technology Benefits*, Research Triangle Park: ISA, 2006 (ISBN: 978-1-55617-905-1).

2. Kirk-Othmer, *Encyclopedia of Chemical Technology*, 3rd edition, Vol. 14, pg. 361, John Wiley & Sons, 1981.

3. Moffat, W., Walmsley, M.R.W., "Improving Energy Efficiency of a Lime Kiln," proceedings of Joint SCENZ/FEANZ/SMNZI Conference, pp. 17-24, Hamilton, July 2004.

4. Chmelyk, T., "An Integrated Approach to Model Predictive Control of an Industrial *Lime Kiln*," Paper presented at the American Control Conference, Anchorage, Alaska 2002.

Appendix A

A web site has been established to enable you to complete the workshops defined in the chapters on process characterization, single-loop and multi-loop control techniques, Model Predictive Control and process applications. In addition, some information is provided on the web site that may be helpful in exploring basic and advanced control techniques.

All you need to access and use the web site is a web browser. Information on how to access the web site and to find the workshops and application exercises is contained in the following section.

A.1   Accessing the Web Site

Using your web browser, go to http://www.controlloopfoundation.com. The workshops that go with Control Loop Foundation are displayed on the home page, as shown in Figure A-1.

To access the application examples contained in Control Loop Foundation, glide the mouse pointer over Applications and click the mouse over the application name.

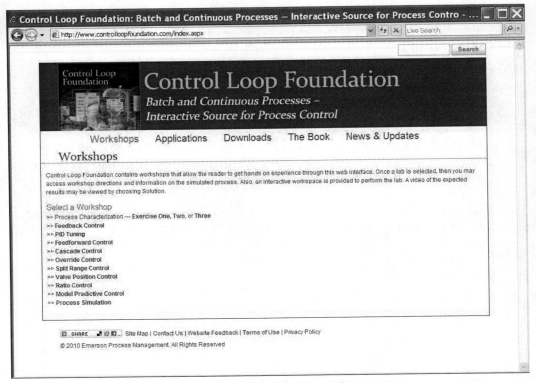

Figure A-1. Web Site Home Page

When the Applications area is selected, the interface will change to show the application examples from this book, as illustrated in Figure A-2.

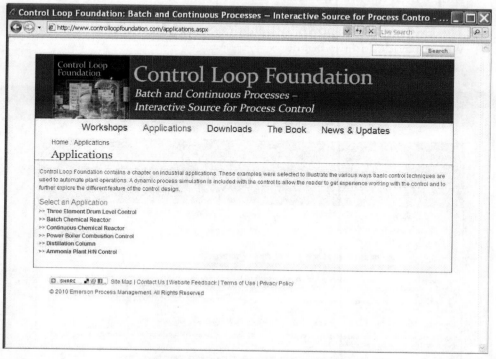

Figure A-2. Application Selections

When a selection is made from the Workshops or Applications interface, the web page will change to give information on the selected workshop or application. For example, if Feedback Control is selected under Workshops, then the interface for this workshop is shown in Figure A-3.

The Exercise tab is shown by default and duplicates the directions that were provided in the book.

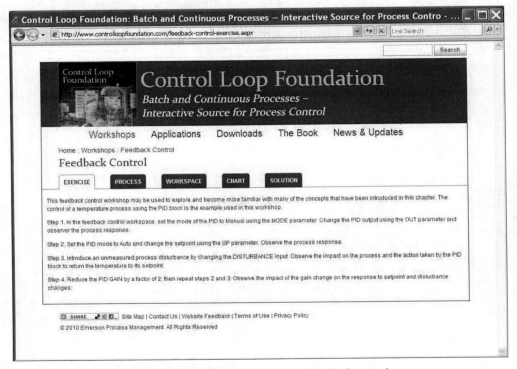

Figure A-3. Workshop – Feedback Control

To see the process that is simulated in a workshop or application, the Process tab may be selected. In response, the bottom portion of the screen will change to show a picture of the process, as illustrated in Figure A-4.

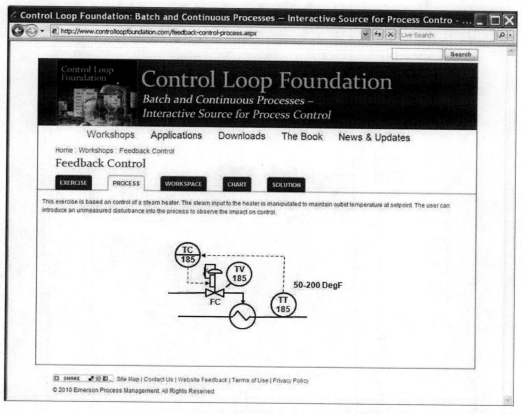

Figure A-4. Process – Feedback Control

With the selection of the Workspace tab, the lower portion of the screen changes to show an on-line view of the function blocks used in the control strategy, as illustrated in Figure A-5.

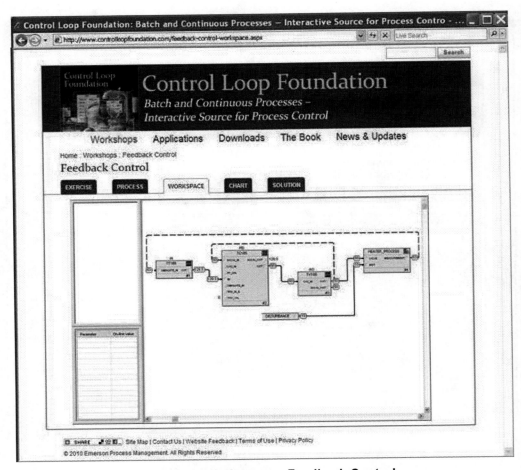

Figure A-5. Workspace – Feedback Control

To change a function block parameter such as the output, setpoint, or mode of a block, click on the block. In response, an indication of the selected block is provided. Also, the parameters associated with the block are displayed in the lower left-hand pane, as illustrated in Figure A-6.

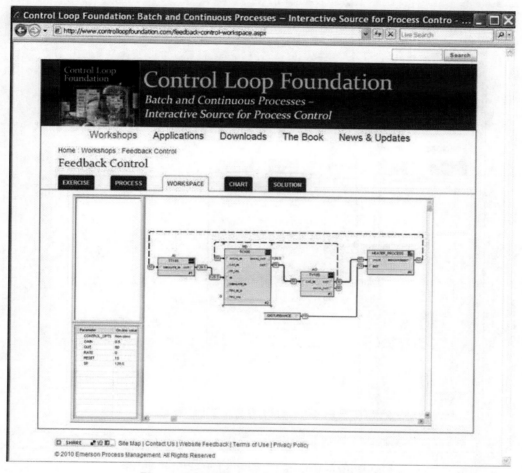

Figure A-6. Selecting a Function Block

To change the value of a parameter, double click on the parameter name in the left pane. In response, a dialog box is displayed that allows the parameter value to be changed, as illustrated in Figure A-7.

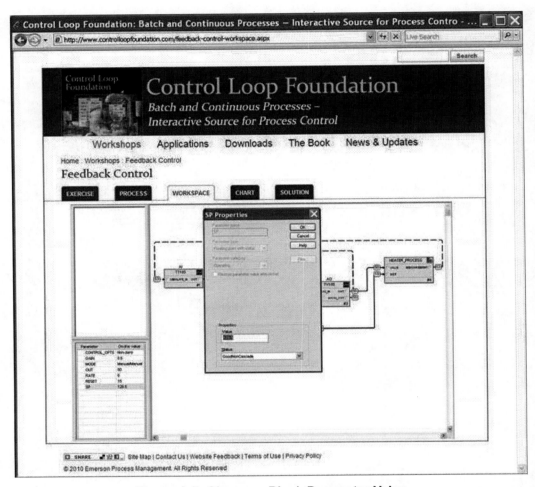

Figure A-7. Change a Block Parameter Value

In some cases, a module parameter is used to allow disturbances to be introduced into the process simulation. To modify a parameter contained in a module, such as the DISTURBANCE parameter in Figure A-7, it is only necessary to double click on the parameter. In response, a dialog box is displayed that may be used to change the parameter value, as illustrated in Figure A-8 for the Disturbance parameter.

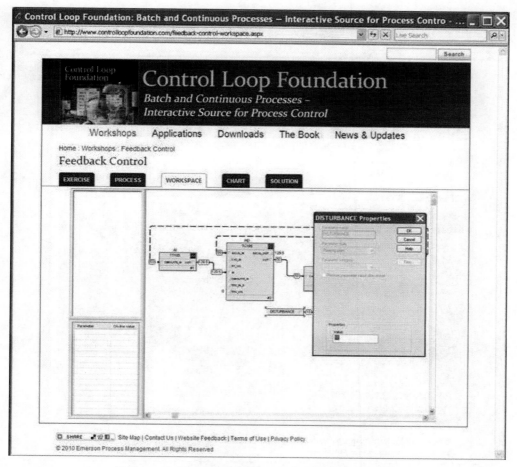

Figure A-8. Change a Module Parameter Value

The impact of a change in a block or module parameter may be observed by examining the dynamic values of the block input and output parameters that are constantly updated based on the simulated process. However, it is often helpful to view a trend of a parameter to better judge the impact of a parameter change. Selecting the Chart tab will cause a trend of selected parameters to be displayed in the lower portion of the screen, as illustrated in Figure A-9.

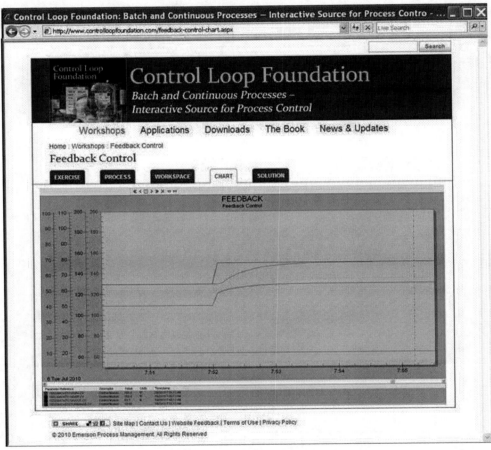

Figure A-9. Chart – Feedback Control

Selecting the Solution tab will cause a video showing the exercise solution to be displayed in the lower portion of the screen, as shown in Figure A-10. In the video, each step of the exercise is shown and discussed.

Using the controls provided in the video area, it is possible to regulate the sound level and to start, stop, and replay the video.

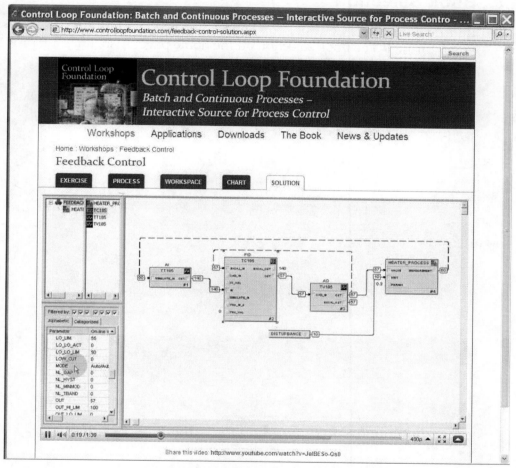

Figure A-10. Solution – Feedback Control

A.2 Download Selection

On the web site home page, a Download selection is provided to allow configuration files for the process simulation and the control modules to be downloaded. These .fhx files may be imported and used with a DeltaV control system or used with the DeltaV Simulate product.

A.3 Book Selection

Information on this Control Loop Foundation book and its authors is provided through The Book selection on the home page. Also, links are provided to allow the book to be purchased over the Web.

Glossary of Terms

Accuracy – The ability of a measurement to match the actual value being measured.

Actual Mode – The mode of operation that may be achieved based on the target mode and the status of the PID block inputs.

Aliasing – Measurement distortion introduced by under sampling the analog value.

Batch Process – Process that receives materials and processes them into an intermediate or final product using a series of discrete steps.

Bumpless Transfer – Method designed to minimize the change in setpoint value on a transition in block mode.

Calibrating a Transmitter – Procedure of adjusting a transmitter to accurately reflect reference input values or a known differential pressure or other process measurement.

Cascade Inner, Secondary, or Slave Loop – PID in a cascade strategy that manipulates the process input.

Cascade Outer, Primary, or Master Loop – PID associated with the final output of the processes making up a cascade.

Cascade Intermediate Loops – PIDs positioned between the outer and inner loops of a cascade.

Closed Loop Control – Automatic regulation of a process input based on a measurement of process output.

Commissioning – The work required to bring a system to the point where it can be used.

Constraint Limit – Value that a constraint parameter must not exceed for proper operation of the process.

Constraint Output (constraint parameter) – Process output that must be maintained within an operating range.

Continuous Process – A *process* that continuously receives raw materials and continuously processes them into an intermediate or final product.

Control Faceplate – Display screen that mimics the manner in which information was provided on an analog controller.

Control Loop – One segment of a process control system.

Control Module – Container of measurement, calculations, and control implemented as function blocks. A module may contain other control modules.

Control Panel – Wall containing controllers, annunciators, and other components used by an operator.

Control Robustness – A measure of how well control copes with differences between the true process gain and dynamics and that assumed in setting PID tuning.

Controlled Output (controlled parameter) – Process output that is to be maintained at a desired value by adjustment of process input(s).

Control Room – Room in which the plant Operator works.

Control System – A component, or system of components functioning as a unit, which is activated either manually or automatically to establish or maintain process performance within specification limits.

Controller Boards – Contains the memory and processor used to perform measurement, calculation, and control functions.

Critically Damped Response – Controlled parameter is restored in minimum time without overshooting the setpoint value. While minimizing response time, unstable control may be observed with changing operating conditions that impact process gain or response time.

Damping – Measurement filtering supported by a transmitter.

Dead Band – The range through which an input signal may be varied, upon reversal of direction, without initiating an observable change in output signal.

Disturbance Input – Process input, other than the manipulated input, which affects the controlled parameter.

Distributed Control System – Control system made up of components interconnected through digital communication networks.

Duty Cycle Control – Regulation of a process input by varying the percentage of the time the process input is turned on over a specific period of time known as the duty cycle.

Dynamic Compensation Network – Network included in the feedforward path to account for differences between the process response to changes in a disturbance input and changes in the manipulated input.

Electronic Control System – System that relies on electric power for the operation of process measurement, control and actuators.

Electro-pneumatic transducers – Device that converts current or voltage input signals to proportional output pressures.

External Reset Feedback – Calculation of the PID integral contribution based on a measurement of the manipulated process input.

Equipment Turn Down – Ratio of the maximum to minimum flow rate supported by a piece of equipment.

Failsafe Position – Position the valve will revert to on loss of connection to the control system or loss of actuator power supply.

Flashing – Change from a high pressure liquid to a low pressure liquid/gas mixture.

Four-Wire Device – Field device that requires a local source of power in addition to the pair of wires used for communication with the control system.

Function Block – A logical processing unit of software consisting of one or more input and output parameters.

Gain Margin – Change in process gain that would be required for an unstable closed loop response.

Grab Sample – Manually acquired sample of process product or feed material taken for lab analysis.

Graphic Display – Display containing a pictorial representation of process equipment and piping along with measurement values at the point they are made.

Hysteresis – The difference in the input signals required to produce the same output during a single cycle of input signal when cycled at a rate below that at which dynamic effects are important.

Increase Open/Increase Close Option– Control system feature that supports valve failsafe setup and permits the operator and control system to work in terms of implied valve position.

Increase/Decrease Control – Regulation of a motorized valve actuator through the use of discrete contacts to run the motor in a forward or reverse direction.

Increase/Decrease Selection – Control system feature that accounts for valve failsafe setup and permits the operator and control system to work in terms of implied valve position.

In-situ Analyzer – Analyzer that provides a stream property or composition measurement by using a sensor directly immersed in the liquid or gas process stream.

Installed Characteristics – Manner in which flow rate changes with valve stem position as installed in the process.

Integrating Gain – Rate of change in the output of an integrating process per percent change in input per second.

I/O (Input/Output) Boards – Contains electronics to convert the field measurements into a digital value that could be used by the controller and to convert digital outputs to electronic signals provided to field devices.

Loop Diagram – Drawing that shows field device installation details including wiring and the junction box (if one is used) that connects the field device to the control system.

Manipulated input (manipulated parameter) – Process input that is adjusted to maintain the controlled parameter at the setpoint.

Manual Control – Plant operator adjustment of a process input.

Media – A liquid or gel designed to support the growth of microorganisms or cells.

MISO Controller – Control based on the use of multiple process inputs and outputs to determine a manipulated process input.

Monotronic – Either entirely nonincreasing or nondecreasing.

Multi-loop Controller – Device for interfacing to multiple field measurements and actuators for the purpose of calculation and control.

Normal Mode – The mode the PID is expected to be in for normal operation.

Other Input – Process input that has no impact on controlled or constraint outputs.

Other Output – Process output other than controlled or constraint outputs.

Output Deadband – Smallest percent change in direction of control system output that is required to change the value of the controlled parameter.

Output Resolution – Smallest percent change in control system output that will be seen in the controlled parameter value.

Over-Damped Response – Controlled parameter is gradually restored without overshooting the setpoint value. In most cases, an over-damped response provides best response for varying operating conditions that impact process gain and response time.

Pairing of Control Loops – Selection of controlled and manipulated parameters to minimize interaction between control loops.

Permitted Mode – The mode(s) available to the operator for a given application.

Phase Margin – Added phase shift through the process that would be required for an unstable closed loop response.

PID Tuning Parameters – Parameters of the PID algorithm that may be used to compensate for process gain and dynamic response and to tailor the control response to changes in setpoint and to process disturbance inputs.

Piping and Instrumentation Diagram – Drawing that shows the instrumentation and piping details for plant equipment.

Plant Operator – Person responsible for the minute-to-minute operation of one or more process areas within a plant.

Pneumatic Control System – System that relies on a pressurized air supply for the operation of process measurement, control, and actuation devices.

Pressure Head – Pressure induced by a given height (or depth) of a liquid column at its bottom end.

Process – Specific equipment configuration (in a manufacturing plant) which acts upon inputs to produce outputs.

Process Action – Mechanism for connecting or coordinating the functions of different PID components.

Process Area – A functional grouping of equipment within a plant.

Process Flow Diagram – Drawing that shows the general flow between major pieces of equipment of a plant and the expected operating conditions at target production rate.

Process Deadtime – For a step change in the process input, process deadtime is defined as the time measured from the input change until when the first effects of the change are detected in the process output.

Process Gain – For a step change in a process input, the process gain is defined as the change in process output divided by the change in process input. The change in output should be determined after the process has fully responded to the input change.

Process Time Constant – For a step change in process input, the time measured from the first detectable change in output until it reaches 63% of its final change in value.

Process Units – Pieces of equipment in a process area that are similar in construction and function.

Pseudo Random – A series of changes that appears to be random but are in fact generated according to some prearranged sequence.

Rack Room – Room designed to house control system components.

Rangeability – The ratio of the maximum full-scale range to the minimum full-scale range of the flowmeter.

Remote Seal – Flush-mounted diaphragm installed at the process vessel and connected to the differential pressure transmitter using a capillary tube filled with fluid.

Repeatability – The closeness of agreement among a number of consecutive measurements of outputs for the same input.

Reset Windup – Continued PID accumulation of reset contribution when further changes in the PID output will have no impact on the process.

Resolution – The ability to detect and faithfully indicate small changes.

Rotary Valve – A type of valve in which the rotation of a passage or passages in a transverse plug regulates the flow of liquid or gas through the attached pipes.

Sampling Analyzer – Analyzer that includes a sampling system to condition a sample of the gas or liquid stream then measured by the sensor.

Self-Documenting – The automatic creation of documents that follow defined conventions for naming and structure.

Self-Regulating – The ability or tendency of the process to maintain internal equilibrium.

Setpoint – Value at which the controlled parameter is to be maintained by the control system.

Single-Loop Control – The manual or automatic adjustment of a single manipulated parameter to maintain the control parameter at a target value.

Slurry – A suspension of insoluble particles in a liquid.

Span – The difference between the minimum and maximum output signals of a pressure sensor.

Step Change – A sudden change from one steady-state value to another.

Superheated Steam – Steam whose temperature, at any given pressure, is higher than water's boiling point.

System Configuration – Specification of measurement, calculation, control and display functions of a control system.

Tag Number – Unique identifier that is assigned to a field device.

Target Mode – The mode of operation requested by the operator.

Terminator – An impedance-matching module used at or near each end of a fieldbus segment.

Terminations – Wiring junctions for field wiring and signal processing to provide an interface to I/O boards.

Transmitter – Field device consisting of a housing, electronics, and sensing element that is used to measure process operating conditions and to communicate these values to the process control system.

Transport Delay – Time required for a liquid, gas, or solid material flow to move from one point to another through the process.

Turndown – Ratio of the maximum to the minimum flow rate that may be measured by a transmitter.

Two-Wire Device – Field device that uses a single pair of wires for power and also to support communication between the device and the control system.

Ultimate Gain – Proportional gain required to achieve sustained oscillation of the controlled and manipulated parameter (with integral and derivative gain set to zero).

Ultimate Period – Period of oscillation achieved for sustained oscillation when the proportional gain is set to the ultimate gain (with integral and derivative gain set to zero).

Underdamped Response – Controlled parameter overshoots the setpoint value but eventually settles at the setpoint value. May minimize time to get back to setpoint but at the expense of stability and overshoot to setpoint and load disturbances.

Underspecified – To give insufficient, or insufficiently precise, information to specify completely.

Unstable Response – Controlled and manipulated parameters exhibit an oscillatory response that builds in amplitude until the manipulated parameter reaches its output limits.

Valve Actuator – Device that produces linear or rotary motion in a valve using a pneumatic, electric, or hydraulic source of power.

Valve Bonnet – Component that provides a leakproof closure for the valve and contains a packing used to maintain a seal between the bonnet and the stem.

Valve Characteristics – Manner in which flow through the valve changes as a function of stem position for a constant differential pressure across the valve.

Valve Positioner – Device used to regulate the valve actuator to maintain the stem position at a target value.

Valve Seat – Replaceable surface used to provide consistent flow shutoff.

Valve Stem – Shaft used to move the variable restriction and thus change the flow through a valve.

Zero Cutoff – Flow rate below which the transmitter indicates a value of zero for the flow measurement.

Index